Pedro Maria-Sánchez

**Neuronal Risk Assessment System
for Construction Projects**

AUS FORSCHUNG UND PRAXIS
BAND 4

Dr.-Ing. Pedro Maria-Sánchez

Neuronal Risk Assessment System for Construction Projects

Schriftenreihe des Instituts für Baubetriebswesen
der Technischen Universität Dresden
Herausgegeben von Prof. Dr.-Ing. R. Schach

Bibliografische Information der Deutschen Bibliothek
Die Deutsche Bibliothek verzeichnet diese Publikation
in der Deutschen Nationalbibliografie;
detaillierte bibliografische Daten sind im Internet über
http://dnb.ddb.de abrufbar.

Bibliographic Information published by Die Deutsche Bibliothek
Die Deutsche Bibliothek lists this publication
in the Deutsche Nationalbibliografie;
detailed bibliographic data are available in the internet at
http://dnb.ddb.de.

Der vorliegende Band 4 der Schriftenreihe des Instituts für Baubetriebswesen wurde durch die Fakultät Bauingenieurwesen der Technischen Universität Dresden als Dissertationsschrift *Neuronal Risk Assessment System for Construction Projects* angenommen und am 20.01.2004 in Dresden verteidigt.

ISBN 3-8169-2483-2

Bei der Erstellung des Buches wurde mit großer Sorgfalt vorgegangen; trotzdem können Fehler nicht vollständig ausgeschlossen werden. Verlag und Autoren können für fehlerhafte Angaben und deren Folgen weder eine juristische Verantwortung noch irgendeine Haftung übernehmen. Für Verbesserungsvorschläge und Hinweise auf Fehler sind Verlag und Autoren dankbar.

© 2005 by expert verlag, Wankelstraße 13, D-71272 Renningen
Tel.: +49 (0) 7159-9265-0, Fax: +49 (0) 7159-9265-20
E-Mail: expert@expertverlag.de, Internet: www.expertverlag.de
http://www.expertverlag.de
Alle Rechte vorbehalten
Printed in Germany

Das Werk einschließlich aller seiner Teile ist urheberrechtlich geschützt. Jede Verwertung außerhalb der engen Grenzen des Urheberrechtsgesetzes ist ohne Zustimmung des Verlags unzulässig und strafbar. Dies gilt insbesondere für Vervielfältigungen, Übersetzungen, Mikroverfilmungen und die Einspeicherung und Verarbeitung in elektronischen Systemen.

Preface of the publisher

Risk management is ever since human constructs an essential assignment because risks related to construction are in many regards different to risks in daily life. The builder-owner has to carry many risks. Especially he has to define the right program and to oblige the right consultants, good designers and efficient contractors. He finally has to coordinate all of them. Buying a new suit, which you do not like any more after a few weeks, is an acceptable risk. Making a heavy accident with a brand new car may be a risk, which may cost you several months of income. Nevertheless, it is possible to get insurance for this risk. Making wrong decisions by building a house can ruin you for a lifetime.

Contractors in construction industry face different risks than other entrepreneurs and in many cases, they are higher. Therefore, in all countries systems have been developed to deal with these risks. Setting up and guiding a construction company remains in many regards risky but gives as well many chances. Therefore, construction managers have to build up risk management systems to handle and to limit the risks.

It has been proven and it is generally accepted in the construction industry, that in most cases the risks, which finally came in and caused high losses were known already before signing the contract. Therefore efficient risk management systems have to function not only during the actual construction process but already during the tender process.

In many cases, risk management systems imbedded in construction companies are just qualitative, often without taking consideration of the results of dealing with similar risks in completed projects.

It is the merit of Dr.-Ing. Pedro Maria Sánchez to develop a quantitative method based on Artificial Neuronal Networks during his time at the Technische Universiät Dresden to evaluate risks before making an offering. It is based on the German methods to calculate offers but it should be possible without serious problems to adapt this model to other procedures of estimating. However, this work is also a good source of informations for people, who just want to get acquaint with risk management in construction industry.

Getting part as volume 4 in the series "Aus Forschung und Praxis" published by the expert verlag this work will bring hopefully important ideas in risk management to many interested persons.

Dresden, January 2005   Univ.-Prof. Dr.-Ing. Rainer Schach

Preface of the author

I am indebted to many people who have provided me with guidance, support and also who gave me helpful comments.

My thanks go first to my personal tutor and "Doktor Vater" Prof. Dr.-Ing. Rainer Schach, director of the Institut für Baubetriebswesen. He has been a strong support along my entire research work at the institute, but also since my first day in Germany and not only regarding with my research challenges, also with many others external situations. He was at every time and moment able to listen, to help and to guide me. I am very pleased with the opportunity of working together; his guidance throughout my whole time in Dresden was very valuable.

To my colleagues of the Institut für Baubetriebswesen, who all the time where able to help me and offer their kind companion all this time. Frau Radloff, Frau Scharmer, Prof. Schindler, Prof. Jehle, Dr. Sperling, Herr Böhme, Thomas Heilfort, Ingo Flemming, Volkhard Gürtler, Jens Otto, Stefan Seyffert, Amira Khneisseh, Aiman Ismail and René Naumann-Jährig. Many thanks for been my colleagues and I hope to work in the future together again.

To Dr. Michael Mawdesley, who was my tutor in my MSc course and now he kindly accepted to be the first expert witness for my doctorate work. His valuable knowledge in many topics of construction management and especially in risk management makes me to feel very proud to have him as my reviser.

To Prof. Bargstädt, from the Bauhaus-Universität Weimar who offered very kindly to be my second expert witness after my presentation at his institute.

To Dr.-Ing. Flach and his team of the Neuronal Research Group of the Technische Universität Dresden (Fakultät Informatik) for guiding me in the topic of artificial neural networks.

Special thanks are for my parents, my mother Martha and my father Pedro who gave me the life and then, they invested on me to be a better human being for the world's society. As well, these especial regards are for my two brothers; Juan Carlos and Pablo Alejandro.

As well, distinguished are dedicated to the Romek family from Skawina, Poland. This whole work, is dedicated to God and to the inspiration, motivation and energy of my life, my wife Dorota Magdalena María Sánchez.

Contents

List of Tables..	V
List of Figures..	VI
Glossary...	VIII

1 Introduction.. 1
    1.1 Concept definition.. 1
    1.2 Motivation... 1
    1.3 Structure of the dissertation... 2

2 What is a Risk - What is a Chance.. 5
    2.1 Risk and chance definitions... 5
    2.2 Other definitions.. 6
    2.3 Classification... 7
    2.4 Risks and Chances in the Construction Industry........................... 10
    2.5 Formal Risk Management... 11
    2.6 Project Risk Management and Enterprise Risk Management....... 13
    2.7 Methods for the Evaluation of Risks... 15

3 Risks and Chances in Construction Companies................................ 21
    3.1 Handling of risks in construction companies................................ 21
    3.2 Specific risks in construction companies...................................... 24
    3.3 Risk management systems in construction companies................. 26
    3.4 Risk management and organisation... 29
    3.5 Company management tasks... 30
    3.6 Risk management as a part of the company management............ 36
    3.7 Systematic solutions... 38

4 Risk Management Methods in the Construction Industry................ 41
    4.1 Net Present Value-at-Risk Method in Infrastructure Project
        Investment Evaluation (Sudong Ye and Robert Tiong).................. 41
        4.1.1 Introduction, description and discussion............................. 41
        4.1.2 Conclusion.. 42
    4.2 Value at Risk (Linsmeier and Pearson)... 43
        4.2.1 Introduction.. 43
        4.2.2 Value at Risk methodologies... 44
        4.2.3 Conclusion.. 46
    4.3 Genesis of Management Confidence Technique (Jaafari)........... 47
        4.3.1 Introduction.. 47
        4.3.2 Objective and requirements... 47
        4.3.3 Conclusion.. 51
    4.4 Estimating using risk analysis (ERA) (Mak and Picken).............. 52
        4.4.1 Introduction.. 52
        4.4.2 Advantages of ERA.. 54
        4.4.3 Conclusion.. 55

  4.5 Risk and Opportunity Analysis Device (ROAD) (Link)........................ 55
    4.5.1 Introduction.. 55
    4.5.2 Project phases.. 56
    4.5.3 Conclusion... 58
  4.6 Evaluation of the risk management methods................................ 59
5 Artificial Neural Networks... **61**
  5.1 What is an artificial neural network?.. 61
  5.2 The structure of the artificial neural network................................. 62
    5.2.1 Micro structure.. 62
    5.2.2 Meso structure.. 63
    5.2.3 Macro structure... 64
  5.3 Learning methodologies.. 64
    5.3.1 Supervised learning... 65
    5.3.2 Unsupervised learning... 66
  5.4 Neural networks transfer functions... 66
    5.4.1 Log-Sigmoid transfer function..................................... 67
    5.4.2 Tan-Sigmoid transfer function..................................... 67
    5.4.3 Linear transfer function (Purelin)................................. 68
  5.5 Work and functions of the artificial neural networks...................... 69
  5.6 Backpropagation algorithm.. 71
  5.7 Example of the backpropagation algorithm.................................. 74
    5.7.1 Forward Propagation... 74
    5.7.2 Backpropagation... 77
  5.8 The Matlab programme.. 81
6 Applications of Artificial Neural Networks in Civil Engineering............... **83**
  6.1 Modelling cost-flow forecasting for water pipeline projects using
    neural networks (A. H. Boussabaine, R. Thomas & T.M.S. Elhag)......... 83
    6.1.1 Project outline.. 83
    6.1.2 Conclusion... 83
  6.2 Application of artificial neural network to forecast construction duration
    of buildings at the pre-design stage (S. Bhokha & S. O. Ogunlana)....... 84
    6.2.1 Project outline.. 84
    6.2.2 Conclusion... 84
  6.3 Neural network model for contractor's prequalification for local
    authority projects (F. Khosrowshahi)... 86
    6.3.1 Project outline.. 86
    6.3.2 Conclusion... 86
  6.4 Artificial Neural Network for Measuring Organizational Effectiveness
    (Sinha and Mc Kim).. 87
    6.4.1 Project outline.. 87
    6.4.2 Conclusion... 87
  6.5 River Stage Forecasting in Bangladesh: Neural Network Approach
    (Liong et al).. 88

6.5.1 Project outline..	88
6.5.2 Conclusion...	88
6.6 Construction Labour Productivity Modelling with Neural Networks (Sonmez et al)	89
6.6.1 Project outline...	89
6.6.2 Conclusion...	89
6.7 General conclusions..	90
7 Theoretical Development of the Neuronal-Risk-Assessment System............	**91**
7.1 Neuronal-Risk-Assessment System Definition......................................	91
7.2 Objectives of the Neuronal-Risk-Assessment System.............................	91
7.3 Working-plan of the Neuronal-Risk-Assessment System........................	92
7.4 Actual Situation of the Problem: What is currently available?..................	93
7.5 Structure of the Neuronal-Risk-Assessment System...............................	96
7.5.1 Risk management...	97
7.5.2 Artificial intelligence..	99
7.5.3 Calculation of construction prices..	100
7.6 Ranking scales of the Neuronal-Risk-Assessment System.......................	103
7.7 Risk-Factors creation...	105
7.7.1 Introduction..	105
7.7.2 Risk-Factors development..	107
7.7.3 Specific Risk-Factors..	107
7.8 Data analysis...	112
7.8.1 Total planned costs, total actual costs and Risk-Factors integration...	113
7.8.2 Risk-Factors and Total-Risk analysis and assessment value......	114
7.8.3 Risk Evaluation Form and artificial neural network creation.....	118
7.9 Neuronal-Risk-Assessment System data flow.....................................	121
7.10 Inputs and outputs model representation...	124
7.11 Analysis of the expected results..	126
8 Empirical Implementation of the Neuronal-Risk-Assessment System (NRAS)..	**129**
8.1 Introduction...	129
8.2 Risk-Factors assessment...	130
8.2.1 Meaning of the Risk-Factors as risk profiles...........................	130
8.3 Total-Risk assessment..	133
8.4 ANN structure..	135
8.5 Testing the NRAS...	137
8.5.1 Testing procedure of the NRAS..	137
8.5.2 Backpropagation network performance at the training phase....	139
8.5.3 NRAS results..	140
8.5.4 The meaning and impact of the NRAS results.........................	143
8.5.5 Risk management evaluation...	152
8.5.6 Conclusions..	153

9 Conclusion..	**155**
9.1 Conclusions..	155
9.2 Limitations...	155
9.3 General conclusions...	156
9.4 Suggestions for further work.....................................	157
Bibliography..	159
APPENDIX I...	165
APPENDIX II..	173

List of Tables

Table 2.1	Comparison of the methods for evaluating the risks...............	19
Table 3.1	Decision, regulation and control risks....................................	22
Table 3.2	Project specific risks..	24
Table 3.3	Risk management methods and systems comparison...............	26
Table 3.4	Steps in the CSA-Q850 Risk Management Decision Process, detailed version, from CSA...	37
Table 4.1	Comparison of Value at Risk Methodologies (Linsmeier and Pearson)...	45
Table 4.2	Scale of Score for each Constraint (Jaafari)...........................	48
Table 4.3	Suggested Initial List of Project-Related Constraints (Jaafari).......	50
Table 4.4	Suggested Initial List of Management-Related Constraints (Jaafari)..	50
Table 4.5	Suggested Initial List of Environment-Related Constraints (Jaafari)..	51
Table 4.6	Relationship between Risk Allowance and Risk Category ERA (Mak and Picken)...	53
Table 4.7	Weighting of probabilities of subjective estimations related to verbal expressions (Link)...	57
Table 4.8	Possible probability distributions of ROAD...........................	58
Table 4.9	Evaluation of the risk management methods...........................	60
Table 7.1	Calculation process of the "Gross Offer Amount" (AS brutto)......	95
Table 7.2	Risk-Factors assessment framework.......................................	108
Table 7.3	Risk-Factor Assessment framework example........................	117
Table 8.1	Risk-Factors scores...	131
Table 8.2	Planned costs...	134
Table 8.3	Actual costs...	135
Table 8.4	Total-Risk with NRAS(E) using network BP10......................	145
Table 8.5	Total-Risk with the actual process...	146
Table 8.6	Risk management results with the NRAS(E).........................	152
Table 8.7	Risk management results with the actual process..................	153

List of Figures

Figure 2.1	Project activities...	5
Figure 2.2	Total risk classification..	8
Figure 2.3	Risk categories (Keitsch 2000)..	9
Figure 2.4	The Formal Risk Management Process................................	11
Figure 2.5	Risk Management Flow Chart (Mawdesley 1997)................	12
Figure 2.6	Project and Enterprise Risk Management (Aon Corporations 1999)...	14
Figure 3.1	Risk develop along the project phases (Bauch)...................	21
Figure 3.2	Process of a Risk-Management conception (Form and Diederichs 2001)..	23
Figure 3.3	Risk Management Environment in the Construction Industry........	25
Figure 3.4	Company Environment within Risks and Chances..............	25
Figure 3.5	Interactive Risk Management Process (PwC)......................	28
Figure 3.6	Components of the planning phase (Bauch)........................	29
Figure 3.7	Organisation of risk management as a rule-cycle (PwC)......	30
Figure 3.8	Risk Management and the Companies (Teji 2000)..............	32
Figure 3.9	Risk Management process..	33
Figure 3.10	Risk Management Overview (AS/NZS 4360)......................	34
Figure 3.11	Risk Management process (AS/NZS 4360).........................	35
Figure 3.12	Business Risk Management Process (BRMP) (Teji)............	38
Figure 4.1	Calculation of NPV-at-Risk and Confidence Level Based on Cumulative Distribution Function (Sudong Ye and Robert Tiong)..	42
Figure 4.2	Histogram of Hypothetical Daily Mark-to-Market Profits and Losses on a Forward Contract (Linsmeier and Pearson).........	44
Figure 4.3	Relationship between Overall Constraints Value (V) and Overall Failure Risk (ß) (Jaafari).................................	49
Figure 4.4	Example of ERA Worksheet at Sketch Design Stage (Mak and Picken)..	54
Figure 4.5	Risk-Assessment outflow in Crystal Ball (Link)..................	56
Figure 4.6	Arrangement of the project-data (Link)..............................	57
Figure 5.1	Micro structure of ANN (Himanen et al)............................	62
Figure 5.2	Six basic topologies of ANN meso structures (Himanen et al)......	64
Figure 5.3	Perceptron architecture (Himanen et al)..............................	65
Figure 5.4	Log-Sigmoid transfer function (Matlab)..............................	67
Figure 5.5	Tan-Sigmoid transfer function (Matlab)..............................	68
Figure 5.6	Linear transfer function (Purelin) (Matlab)..........................	68
Figure 5.7	Neural network functionality...	69
Figure 5.8	General neural network architecture..................................	70
Figure 5.9	Neuron architecture (Tsoukalas and Uhrig)........................	70
Figure 5.10	Back-propagation neural network.......................................	71
Figure 5.11	Summary of the Widrow-Hoff algorithm (Haykin)..............	72
Figure 5.12	Sample network..	74
Figure 5.13	Outputs of the hidden neurons...	76
Figure 5.14	New weight values..	79
Figure 5.15	Graphical User Interface (GUI)..	82

Figure 7.1	Calculating Risk and Profit by total production costs................	96
Figure 7.2	Structure of the Neuronal-Risk-Assessment System................	98
Figure 7.3	Basic activities of the Neuronal-Risk-Assessment System............	102
Figure 7.4	Risk-Factors (inputs) ranking scale..	104
Figure 7.5	Total-Risk (outputs) ranking scale..	105
Figure 7.6	Purpose of the Risk-Factors..	106
Figure 7.7	Current and NRAS approaches..	114
Figure 7.8	Risk-Factors and Total-Risk relation.....................................	115
Figure 7.9	Risk Evaluation Form...	119
Figure 7.10	ANN structure..	120
Figure 7.11	Data-flow process...	122
Figure 7.12	Use process of the Neuronal-Assessment System with new projects...	124
Figure 7.13	Input-Output relationship..	125
Figure 7.14	ANN vectors...	125
Figure 7.15	Risk behaviour..	127
Figure 7.16	Risk Value with the ARRIBA software...................................	128
Figure 8.1	Risk-Profile for project 2...	131
Figure 8.2	Risk-Profile for project 10...	132
Figure 8.3	Input vector...	136
Figure 8.4	Output vector...	136
Figure 8.5	NRAS testing procedure...	138
Figure 8.6	Backpropagation network performance for NRAS(A)................	139
Figure 8.7	NRAS(A) results..	141
Figure 8.8	NRAS(C) results..	142
Figure 8.9	BP10 errors from model NRAS(E).......................................	144
Figure 8.10	NRAS(E) performance..	147
Figure 8.11	Total-Risk in money terms per project.................................	149
Figure 8.12	Cumulative Total-Risk per approach...................................	150
Figure 9.1	Project's Total-Risk and its causes......................................	158

Glossary

AIA 201	American Institute of Architects Document A201
AS netto	Angebotssumme netto (Net Offer Amount)
AS brutto	Angebotssumme brutto (Gross Offer Amount)
AGK	Allgemeine Geschäftskosten (General business costs)
ANN	Artificial Neural Networks
BGK	Baustellengemeinkosten (General costs of the construction site)
BP	Backpropagation
BRMP	Business Risk Management Process
BGB	Bürgerliches Gesetzbuch (The German Civil Code)
CAD	Computer- aided drafting
EKT	Einzelkosten der Teilleistungen (Single costs of the partial payment)
ERA	Estimating using risk analysis
FIDIC	Federation Internationale des Ingenieurs- Conseils
GDR	Generalized Delta Rule
GUI	Graphical User Interface
HSK	Herstellkosten (Production costs)
ICE	Institution of Civil Engineers
IT	Information Technology
JCT	Standard Form of Building Contract
MCT	Management Confidence Technique
NPV	Net Present Value
NRMS	Neuronal-Risk-Management System
PERT	Program Evaluation and Review Technique
RF	Risk-Factors
ROAD	Risk and Opportunity Analysis Device
V	Overall Constraints Value
VOB	Vergabe- und Vertragsordnung für Bauleistungen (Contracting regulations for building works and supplies)
SB	Selbstkosten (Total Production Costs)
W+G	Wagnis und Gewinn (Uncertainty and Profit)
ß	Overall Failure Risk

1 Introduction

1.1 Concept definition

The objectives of risk management are generally considered to be oriented towards identifying, analysing and responding to risks. However, risk management should be broader than this and become a culture integrated into the company and project environments as a mean for providing safety measures against risks.

Risk management is not a new topic for the construction industry and many organisations employ it as a matter of routine. When they do, they focus mainly on downside risks and use either subjective methods or, in the more 'scientific-based' companies, probabilistic methods (for example Monte Carlo Simulation, Latin-Hypercube and Sensitivity Analysis). Even these latter methods might be considered traditional because they do not offer any new approach to deal with risks. Neither do they offer a solution to one of the big challenges of risk management: the assessment of the value of risks in monetary terms.

The main goal of this research-work is to quantify the cost of the risks involved in construction projects. This is achieved through the construction of a Neuronal-Risk-Assessment System (NRAS) based on artificial neural networks (ANN). The development of this system has required the production of a list of Risk-Factors (RF) to model the behaviour of the risks for any given construction project.

The theoretical development and the practical application of the system are described at the end of this dissertation.

1.2 Motivation

The necessity to formally identify, analyse and respond to risks represents the principal reason and motivation of this work. This research looks to investigate whether it is possible to use artificial neural networks for identifying and evaluating risks in construction projects. The core elements of the system are represented by Artificial Networks and Risk-Factors. The Risk-Factors are the most relevant elements because they provide the data for the neural network. After several discussions with a Dresden-based contractor, a final list of 17 Risk-Factors was created. The main contribution of the Risk-Factors is that they contributed a base from which to evaluate risks for a chosen project.

The benefits of risk management in many construction companies are not fully understand, mainly because its results shown with past approaches were not so tangible. In the past, risk management was mainly referred with losses; however, now the picture of risk management is not only losses; also wins are included. The meaning in practical term of risk management, is that any project offer losses in materials, equipment, human resources. As well, wins can be achieved with finishing the project in time, saving resources while improving productivity, reducing the accidents at the construction site, etc.

The NRAS is a very flexible tool. However, throughout this work the application of the system was concentrated at the operational management level, in order to cover a complete theoretical and practical development, as is portrayed at the end of this work.

At the time for bidding for a project a contractor does not have sufficient information to make a valid judgment about the value of risk. It is common practice in Germany to use a percentage of total cost to allow for risks. This is usually between two and three percent. However, the reality is that risks vary for different types of projects. It is also unrealistic to use the same risk value for projects which differ in the location. Being realistic, it is necessary to assume that every project is unique and that every project offers unique risks.

The NRAS possesses the flexibility, that can be used for different type of projects, by adopting the Risk-Factors for the specific situation. The NRAS is a human intuition approach, because it uses the experience of the project manager to measure the reliability of the system. It is also complemented by the theory of risk management and artificial neural networks.

1.3 Structure of the dissertation

Following this introduction the fundamental definitions and concepts of the risk management process are described and discussed in chapter 2. The meaning of risk management as part of company's structure, and its role in a project are also analysed and commented upon.

A brief history of risk management within the construction industry is presented in chapter 3. The applied background of risk management in companies, projects and processes is described and discussed.

The methodologies that have been used for managing risks in construction projects are contained in chapter 4. A brief but detailed explanation of how these techniques work is included; in addition, the types of results offered are debated.

In chapter 5, an introduction, background, definition and the function ability of artificial neural networks are included. In addition a simple but practical example is developed. Some relevant examples of applications of artificial neural networks in civil engineering are incorporated in chapter 6.

The development of the theoretical framework of the NRAS is described in chapter 7. The aim objectives the structure of the NRAS are covered. Results and further recommendations are also included.

The work done in chapter 7 is used in chapter 8 in order to carry out the training and testing of the NRAS. In this chapter, data from 16 finished projects is used to create the artificial neural network, to train it and finally, to perform the simulation. The final outcome of the practical application of the NRAS provides the Total-Risk values per project in monetary terms.

In chapter 9, the weakness and strength of the NRAS are presented and discussed. In addition; the potential contribution of the NRAS to the practical life of the construction industry is discussed. Further recommendations for improving the system are presented.

2 What is a Risk - What is a Chance

2.1 Risk and chance definitions

It is clear that when we talk about risks and chances, we are talking about losses and benefits. These two elements are deeply involved in every construction process and a full understanding of what they mean is a good first step into the risk management world.

Mawdesley[1] defines risk and chance as:

Risk: The potential of an uncertain outcome of an event to "threaten" a held objective.

Chance: The potential of an uncertain outcome of an event to "promote" a held objective.

As well, we find by Link[2] the following definitions:

Risk: Is the probability that a negative event has in order to occur, the impact of this event affects the original target. The results of this event are negative.

Chance: Is the probability that a positive event has in order to occur, the impact of this event affects the desired target. The results of this event are positive.

Figure 2.1 shows the relationship of the risks and chances as outcomes of the project activities (planned costs - actual costs)

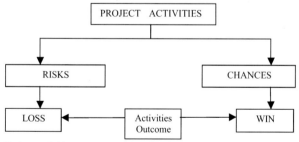

Figure 2.1   Project activities

---

[1] Mawdesley, M., Askew, W., and O'Reilly, M. (1997). Planning and controlling construction projects: The Best Laid Plans.

[2] Link, D. (1999). Risikobewertung von Bauprozessen.

2.2 Other definitions

"Risk" and "Chance" are the only terms which must be defined before risk management can be discussed. Mawdesley[3] provides the following definitions:

Uncertainty: The quality associated with an event which results in an inability to predict its outcome accurately.

Risk Analysis: Is the part of the risk management process that determine the expected values for these uncertainties. This is a task for estimators and economists, whose training and experience enable them to estimate the expected values of the various identified variables and their likely ranges.

Risk Response: This stage of the risk management process reflects the views and policies of the managers involved.

Other relevant definitions indicated by the Australian/New Zealand Standard on Risk management[4] are as follows:

Residual risk: The remaining level of risk after risk treatment measures have been taken.

Risk acceptance: An informed decision to accept the consequences and the likelihood of a particular risk.

Risk analysis: A systematic use of available information to determine how often specified events may occur and the magnitude of their consequences.

Risk assessment: The overall process of risk analysis and risk evaluation.

Risk avoidance: An informed decision not to become involved in a risk situation.

Risk control: The part of risk management which involves the implementation of policies, standards, procedures and physical changes to eliminate or minimise adverse risks.

---

[3] Mawdesley, M., Askew, W., and O'Reilly, M. (1997). Planning and controlling construction projects: The Best Laid Plans.

[4] AS/NZS 4360 (1999). Risk Management.

Risk engineering:	The application of engineering principles and methods to risk management.
Risk evaluation:	The process used to determine risk management priorities by comparing the level of risk against predetermined standards, target risk levels or other criteria.
Risk financing:	The methods applied to fund risk treatment and the financial consequences of risk.
Risk reduction:	A selective application of appropriate techniques and management principles to reduce either likelihood of an occurrence or its consequences, or both.
Risk retention:	Intentionally or unintentionally retaining the responsibility for loss, or financial burden of loss within the organisation.
Risk transfer:	Shifting the responsibility or burden for loss to another party through legislation, contract, insurance or other means. Risk transfer can also refer to shifting a physical risk or part thereof elsewhere.
Risk treatment:	Selection and implementation of appropriate opinions for dealing with risk.

2.3 Classification

Risk events can be classified by their outcome. In other words, depending on how strongly an event affects the main objectives of the project. Figure 2.2 shows a proposal of the classification effect used in this research. This classification of the risk is just a basic example, but is the core of the classification used for the risk in chapter 7.

It is basic, because the theoretical development made in chapter 7 over the NRAS form the central idea of this research; using several RF identified together with different designs of ANN. It is in chapter 8 where the application phase of the NRAS takes part, discussing afterwards its advantages and disadvantages.

The foundation of the total risk classification shown in figure 2.2, is supported by an assessing scale of risk (Method 3: Skewed assessment, Toolbox 4, page 29) of the "A simple guide to controlling risk" CIRIA[5]. The percentage values of the risk impacts are proposed on this work.

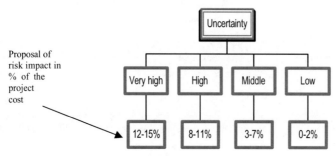

Figure 2.2   Total risk classification

The total risk impact is related directly to the impacts of the risks considered in the assessment of a particular project. Even when the probable value of the impact of the total risk in a project has been evaluated (for example high, 8%), there will be other events (during the execution phase of the project) which might cause an increment on this value; events like earthquakes, political and social riots, weather phenomena. In this case, even if a good evaluation of the project has been made, the total risk will be increased.

The World Bank appraises projects using social, commercial and political criteria. King (1967)[6] provides a large amount of information about the methods used by the World Bank to evaluate the projects. In each case, only the possible negative effects are evaluated in non-economic terms (refer to chapter 7 and 8 where is shown a proposal to quantifying risks in money terms).

According to Keitsch[7], there is also another classification of risks available. Risks can be classified in different categories. For example in political, environmental risks, project risks, company risks and "Act of God risks". In Figure 2.3, three categories of risks are shown. However, it is important to mention that this classification is limited, because other existing types of risks such as social risks and management risks are not included.

---

[5] CIRIA (2002). A simple guide to controlling risk.

[6] King, J. (1967). Economic Development Projects and Their Appraisal: Cases and Principles from the experience of the World Bank.

[7] Keitsch, D. (2000). Risikomanagement.

What is a Risk- What is a Chance 9

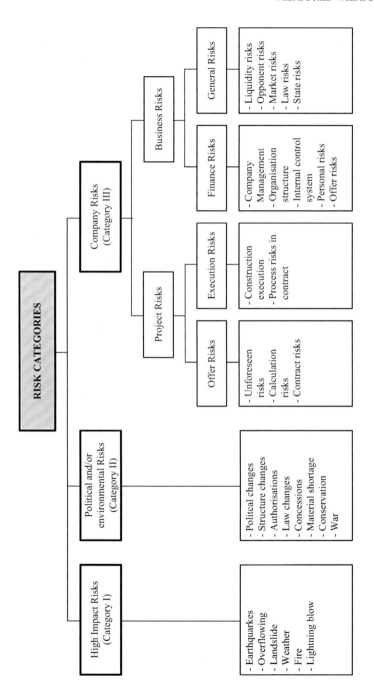

Figure 2.3 Risk categories (Keitsch 2000)

The risks categories shown in figure 2.3 are not wide enough to cover the whole spectrum of the risks which belong to the construction industry. The classification (figure 2.3) shows three main categories, these are: High impact risks, political and/or environmental risks and company risks. Other type of risks which are as important as these mentioned before and that should be included in these main categories are for example: planning risks, management risks, technical and operational risks, safety risks, etc.

The category of High Impact Risks (figure 2.3) is a confusing one. Because it includes a mix of specific risks which can be enlisted in other categories. For example "earthquakes", "weather" and "overflowing" can be better assigned to "acts of God risks", and "fire" to the category of "technical risks".

2.4 Risks and Chances in the Construction Industry

The nature of activities involved in construction means that in each project, there will be risks and chances. In the construction industry a particular or special event (risk and chance) can be explained as: the event where its nature or cause comes from the project activities. In general terms, there are general risks which enclose particular ones. Examples of general risks are: planning risks, operational risks, environmental risks, financial risks, etc.

Among the Risk from Construction: Preparation of a Client's Guide (CIRIA)[8] some examples of particular risks are included:
- design/technical principles unproven
- specification not within bounds of physical possibility
- conflict or incompatibility of design/technical principles
- design brief of project objectives not clearly stated
- unforeseen ground conditions
- adverse performance by contractor on recent projects overlooked
- non performance by contractor concerning quality control procedures
- non availability or scarcity of materials
- problems with supply base e.g. single source
- non availability of alternative materials of components
- unstable project senior management and organisation
- client investment proposal unsound

---

[8] CIRIA (1996). Along the Risk from Construction: Preparation of a Client's Guide.

2.5 Formal Risk Management

Formal Risk Management is a structured approach to administrate (analyse, evaluate and control) risks. It consists of the four concepts show in figure 2.4:

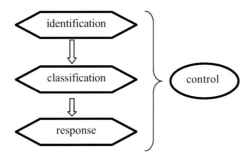

Figure 2.4 The Formal Risk Management Process

Mawdesley[9] defines risk management as "the name given to the formalised process of balancing the risks and chances that a decision may lead to; this involves taking action to produce an acceptable balance between the two".
Mawdesley[9] also provides the Risk Management Flow shown in figure 2.5.

---

[9] Mawdesley, M., Askew, W., and O'Reilly, M. (1997). Planning and controlling construction projects: The Best Laid Plans.

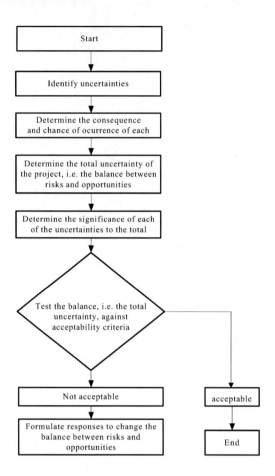

Figure 2.5   Risk Management Flow Chart (Mawdesley 1997)

The Risk Management Process represented in this figure, describes typical overall steps to administrate the risks and opportunities. Although broad, it is enough to follow as a guide to understand the complete process of risk management. The three main components of risk management are: Identification (identify uncertainties), Analysis (determine consequences, determine the total uncertainty and determine the significance of uncertainties and testing the balance) and Response (formulation of responses).

In principle, both risk management processes shown in figures 2.4 and 2.5 enclose the basis of the risk management theory. The main difference between both processes is that the one shown in figure 2.4 is a general approach; its components cover main activities in each one. For example at the identification phase, activities like risk check lists, define the level of risk, define the management level (high, middle or low) etc., are included. In the other hand, in the process shown in figure 2.5, each concept is more defined and it includes a decision stage whether the risk impact is not acceptable or is acceptable. Providing an acceptance of the risk impact, the end of the risk management process and providing a not acceptance the formulation of a response to the risk impact.

2.6   Project Risk Management and Enterprise Risk Management

The working area of the "project risk management", is the project itself. In this case, the risk system is designed and applied under the variables of a project. The goal of the risk system is strongly directed to provide the project with better chances of success. In the other hand, an "enterprise risk management", is fully designed and applied under the environment of the company; tackling risks which are related to the business process.

Hudson[10] defines Enterprise Risk Management as "a way of managing risk and uncertainty in the new economy. It is a new way of thinking about risk. It means aligning and organization strategies, processes, people, technology and knowledge to meet its risk management purpose. Managing risks on an enterprise-wide basis means making an entire organisation aware of risk and equipping everyone to thrive on uncertainty".

Aon Corporations[11] provides the Enterprise Risk Management guide shown in Figure 2.6. This guide consists of seven steps, the first steps (from step one to step three) deal with the identification of risks and the organisation of groups for analysing them. In step four, the risk data is developed. The data include information related to business strategies, risk models, past examples of business failures, benchmarking information, etc. The next step (step 5), is oriented to create a risk inventory; the goal of this risk inventory is to advise the company whether the response to a particular risk is already within its risk inventory. Continuing with step six, this step deals with forecasting the company weakness towards risks. Finally in step seven, a risk management plan should be formulated. This guide is a good attempt to record mainly the identification of risks and also to classify their severity in a qualitative manner. It is a requirement (for its use in construction companies) to implement its framework within several companies in order to analysis and compare the outcomes, advantages and disadvantages towards a Non-Enterprise Risk Management approach.

Enterprise Risk Management is more focused towards an identification of the risks at a company level. Its seven steps described in figure 2.6 are related to the objectives of the

---

[10] Hudson, F.V. (2000). Enterprise-wide Risk Management: Strategies for Linking Risk and Opportunity.

[11] Aon Corporation (4th ed.) (1999). Enterprise risk management, part two.

company, identify key risks, developing a risk data, the environment of the company towards the risk and developing a financial plan. Overall, its structure is acceptable, but is limited to propose any type of response and as well its application as described in figure 2.6 is rather too general.

The usefulness of Enterprise Risk Management within contractors as shown in figure 2.6 is not enough complete. The lack of steps regarding the analysis, assessment, response and control of risks in construction is a strong disadvantage of this guide. However, it should be possible to modify this guide and include more and new relevant steps for managing risks at a company level.

A Do-it-Yourself Guide to Enterprise Risk Management		
STEP ONE. Determine Objectives/Form an Enterprise Risk Management Team - Comprising Professionals with responsibility for risk from throughout the organisation		
STEP TWO. Determine Risk Retention Capacity		
- Earnings Per Share Impact	- Capital Structure	- Rating Agency Issues
STEP THREE. Determine Key Risks Facing the Company		
- Foreign currency	- Distribution	- Liability
- Manufacturing	- Interest Rates	- Revenue
- Human capital	- Regulatory	- Credit
- Commodity	- Intellectual property	- Assets
- Crime	- Geographical	- Political
STEP FOUR. Develop Critical Enterprise Risk Data		
- Loss/exposure Data	- Business Strategies	- Benchmarking Information
- Financial Analysis	- Risk of Ruin Scenario	- Business Frustration Examples
STEP FIVE. Create a Risk Inventory		
- Risk Retention	- Probability	
- Potential Severity	- Financial Impact	
STEP SIX. Plot Risks to Visualise the Company's Total Exposure - Severity/Frequency Low/High		
STEP SEVEN. Use Risk Maps to Develop Optimal Risk Financing Plan - Integrated/Catastrophe - Enterprise/Catastrophe Excess/Contingent Capital Programs - Retained or Self-insured Risks - Aggregate Protection Programs		

Figure 2.6   Project and Enterprise Risk Management (Aon Corporations 1999)

2.7 Methods for the Evaluation of Risks

A wide range of techniques is available for analysing and evaluating the risks of construction projects. Some techniques have a similar basis, some for example are probabilistic, other ones are subjective (mainly in identifying risks), and others have a system based on a learning approach, as used in neuronal networks.

The objectives of risk analysis can be described as:

> ➢ evaluation of the possible consequences of each uncertainty and the probability of occurrence (refer to the Risk-Factors in chapter seven);
> ➢ evaluation of the project's total risk position or the balance between opportunities and risks;
> ➢ estimate the importance of each individual uncertainty to the total result.

A list of the most known methods of risk analysis will follow, together with a short description:

> ➢ **Delphi-Method:** The Delphi method is an exercise in group communication among a panel of geographically dispersed experts (Adler and Ziglio)[12]. The technique allows experts to deal systematically with a complex problem or task. The essence of the technique is fairly straightforward. It comprises of a series of questionnaires sent either by mail or via computerised systems, to a pre-selected group of experts. These questionnaires are designed to elicit and develop individual responses to the problems posed and to enable the experts to refine their views as the group's work progresses in accordance with the assigned task. The main point behind the Delphi method is to overcome the disadvantages of conventional committee action. According to Fowles[13] anonymity, controlled feedback, and statistical response characterize Delphi. The group interaction in Delphi is anonymous, in the sense that comments and forecasts, are not identified as to their originator but are presented to the group in such a way as to suppress any identification.

---

[12] Adler, M., Ziglio, E. (1996). Gazing into the Oracle: The Delpli Method and its Application to Social Policy and Public Health.

[13] Fowles J. (1978). Handbook of Futures Research.

The ideal place of work for the Delphi method can be within the evaluation of risks for example in appraising construction projects from the social point of view. In this case, the project evaluators (government, contractor, financing part and society) can be divided for assessing the project impacts while identifying and analysing the risk factors considered from each side of the project parties.

Ideally, the Delphi method is recommended to use for identifying the main uncertainties involved in this type of evaluation (social) for projects which have a direct impact to the country or community.

The benefits of this technique in one sense can be directed to the government. In this case, the government can receive for example for a proposal of an infrastructure project, different opinions and judgment coming from experts of different background. This information, can be filed into a single document which can be used as a guide for future similar projects.

The main disadvantage of this technique is the resources demanded by implementing this approach. Different groups of evaluators must be coordinated, fed with information, controlled and communicated with; all this demands a considerable amount of money, time and material and equipment resources.

- **Monte Carlo Simulation:** is a technique employing random numbers in order to combine distributed variables (such as the uncertain durations of activities in a network). The technique simulates a project by choosing at random the values for each of the variables and using these to calculate the outcome of the project (Mawdesley)[14].

Monte Carlo Simulation uses variables together with random values for each of them, instead of using variables as the time of construction, the cost of construction, the annual discounted rate, etc. In risk analysis, for example variables which are more directly related to the risks of a chosen project can be used, for example the Risk-Factors developed in chapter seven.

While using these wide spectrum of risks, the probable outcome and impact of these risks which influence the project can be simulated with Monte Carlo. However a considerable disadvantage is, that it requires many runs, and a lot of these runs are required for a project with many risks.

---

[14] Mawdesley, M., Askew, W., and O'Reilly, M. (1997). Planning and controlling construction projects: The Best Laid Plans.

The Monte Carlo simulation is a technique that allows the identification of the critical risks that might occur during the execution of a project. For that reason, it is recommendable to use this method for identifying and assessing the risks at the tendering phase of the project for finding the possible responses for each critical uncertainty or for sharing the risks.

- **Latin-Hypercube Simulation:** Latin Hypercube sampling is generally more precise for producing random samples than conventional Monte Carlo sampling, because the full range of the distribution is sampled more evenly and consistently. Thus, with Latin Hypercube sampling, a smaller number of trials achieves the same accuracy as a larger number of Monte Carlo trials. The added expense of this method is the extra memory required to hold the full sample for each assumption while the simulation runs.

The application proposed for using Latin-Hypercube Simulation in analysing the risks is similar to the one described within the Monte Carlo Simulation. The main difference, is that the probability distribution use for this technique, allows to reduce the number of runs required to obtain acceptable results, while using several risks variables.

This deviation allows to obtain quicker results using the same risk-analysis framework proposed in the Monte Carlo Simulation.

- **PERT (Program Evaluation and Review Technique):** is a technique which attempts to account for the inherent uncertainty of activity durations. It is based on a network model of a project and was developed shortly after the initial development of arrow diagrams (Mawdesley)[15].

Because PERT considers the uncertainty of the activities' durations in a network-based model of a project, its application on the field of risk analysis can be done while considering the most risky activities under a PERT criteria. The most risky activities can be identified for example using the Delphi method, in which perhaps the experts will decide that the most risky ones are those related with highest costs, very specialised activities or activities which demand a wide variables of resources.

Instead of analysing the whole network with PERT, the most critical activities from the risk point of view can be analysed in terms of their duration. An alter-

---

[15] Mawdesley, M., Askew, W., and O'Reilly, M. (1997). Planning and controlling construction projects: The Best Laid Plans.

native approach to combine also with PERT is by using the Critical Path Method to obtain the critical path of a project and analyse with PERT this path.

As disadvantages of this technique can be stated, that for small projects the time invested might not be recompensed by the outcome of the project. The model needs a considerable amount of data to be provided by the project manager (optimistic, pessimistic and normal durations), there is no control recommendation to be applied to the results and the technique is quite old and perhaps needs an update in order to be used with the new theories of risk management.

> **Sensitivity Analysis:** Is a powerful technique commonly used in industry to show the effect each variable has on the outcome of a project. As the name suggests, the technique determines the sensitivity of the project outcome to changes in a variable. The results of this type of analysis are usually presented graphically, one axis representing the percentage change in the variable with the other representing the resulting percentage change in the outcome of the project (Mawdesley)[16].

The application of Sensitivity Analysis in risk analysis is practical and meaningful. Once the identification phase has been done for a particular project, using a prompt list, check list (what can go wrong?) of the risks identified (or the Risk-Factors from chapter seven), several scenarios can be tested while simulating the effect to the outcome of the project caused for a specific variable (risk). For example, the sensitivity of the project towards a variable like weather or unforeseen underground conditions, while giving to these variables different values (in %), this change in % can be in terms of cost, time, discount rate and other more relevant variables.

With this assumption, it is possible to build up several risk scopes for each uncertain variable together with determining each change also in its Net Present Value (NPV) (in money terms and in %). With this framework, the decision maker is able to identify and analyse in detail the most risky variables and consider proper alternatives for solutions.

Table 2.1 shows a comparison of the advantages and disadvantages of the described techniques. It also provides a recommendation about for which kind of projects they should be implemented on.

---

[16] Mawdesley, M., Askew, W., and O'Reilly, M. (1997). Planning and controlling construction projects: The Best Laid Plans.

Method	Advantages	Disadvantages	Suitable for:
Delphi Method	Team orientated, excellent for questionaires, good communication, good for dealing with complex problems	Expensive to apply, takes considerable time, lack of objectivity	Social projects, social evaluations, governments projects
Monte Carlo Simulation	Good record as a simulation method, good obtion to measure results, flexible	Requires good computational equipment, based on experience, migth be expensive, can takes considerable time to design the variables framework	Different types of construction projects, good for construction processes, good for company financial forecasting
Latin-Hypercube Simulation	More precise than Monte Carlo simulation, less time to build the variables framework, good results with variables, flexible	Requires good computational equipment, based on experience, migth be expensive	Different types of construction projects, good for construction processes, good for company financial forecasting
PERT	Good for planning techniques, relative not expensive, good for quantify uncertanty in the project activities, provides an estimate of an uncertain project end date	Limited to planning, takes considerable time to learn, takes considerable time to be implemented, assumes activities are statistically independent, does not provide start and end dates for activities except as in ordinary networks	Good for planning control, for planning the project critical activities, for decision-making over the project activities, good for construction proceses
Sensitivity Analysis	Very well known technique, flexible, fast to be implemented, easy to read and unerstand the results	Too probabilistic method, based on experience	Useful for every type of project, good for company financial forecasting, good for analysing different project alternatives

Table 2.1   Comparison of the methods for evaluating the risks

In other words, table 2.1 is a resume of the most relevant qualification of the techniques mentioned in point 2.7. The table allows the reader to identify the probable resources and requirements necessary to be implemented in each technique. As well, it provides hints for using these techniques in some specific projects, types of evaluation and phases of development of projects.

3 Risks and Chances in Construction Companies

3.1 Handling of risks in construction companies

Construction companies work with risks on a daily basis. From project inception through to project completion, the handling of risks is an essential management task; for doing this, the use of clauses dedicated fully to mitigate, to reduce or to manage (risk responsibilities, risk transfer, risk share) risks are proposed by contractors. These can be used to manage some of the uncertain events, which face a project.

In some conditions of contract (for example FIDIC), various specific risks are mentioned and possible responses included. For example risks like riots, war and earthquakes are frequently included. However the quantification of the impact of these remains is largely unsystematic.

In figure 3.1, Bauch[17] shows how the risk importance develops through time and along the project phases. Based on this figure, it is clear that the risks are more relevant at the planning phase, at the decision phase they reduce considerably and finally at the regulation and control phases the risk importance tends to increase and reduce respectively.

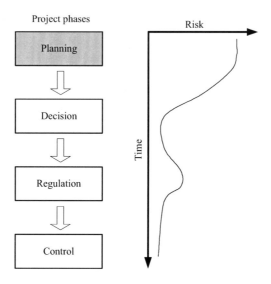

Figure 3.1   Risk develop along the project phases (Bauch)

---

[17] Bauch, U. (1994). Beitrag zur Risikobewertung von Bauprozessen.

However the impact of the risks (as shown in table 3.1) in the later phases can be, in many situations, higher than the impact of the risks at the planning phase. It is not possible, therefore to generalise that the higher risks in construction projects happen at the planning phase.

The risk behaviour shown by Bauch in figure 3.1, is limited to make an emphasis that risks are more dangerous at the planning phase. Nevertheless, the risk impact and as well the role of the risks is also very important in the execution phase of a project. For that reason, with the other phases (decision, regulation and control), the risk importance should be considered to be high. Another phase which is not include in figure 3.1 and is very important, specially when dealing with for example BOT projects is the operational phase, in which the risks can be as well very dangerous.

Decision-making risks	Regulation risks	Control risks
Poor management knowledge	Poor knowledge of construction law	No control system to measure the project performance
Poor communication between the project evaluators	Weak company policies towards subcontractors	No evaluation framework at the bidding phase
Lack of experience in specific type of projects	Contract problems (disruptions)	Poor quality control

Table 3.1   Decision, regulation and control risks

With due regard the handling of risks by contractors, some international construction companies (such as Bechtel) have a standard method (Monte Carlo Simulation) which they always use to evaluate the risks. Others are planning to have one, and others are not in such a procedure. The reality of risks is that risks are dynamic and there is no simple formula or process to quantify all of them. Risks change through time, so the methods currently used must be improved (for example the traditional Monte Carlo Simulation or PERT) or other new ones need to be introduced into the construction industry.

The approach adopted by Form and Diederichs[18] and illustrated in Figure 3.2, shows the different project phases that make up a risk management system for a company. These phases are: Organisation, actual analysis, planned concepts, classification and documentation. For example, in the case of a construction company the organisation phase will deal with the project organisation including the teams that will carry out the project both at the office and at the construction site. In this phase the tasks and responsibilities to be performed by each team or company department will also be defined. The structure (or-

---

[18] Form, S., and Diederichs, M. (2001). Risikomanagement.

ganisation) to be use within the project is discussed and the projects committees are named.

In the actual analysis, the goal is to collect information of how the company has been involved in managing risks and also to reduce the risk inventory of the company. In other words, the current practices of the company towards risks are analysed. At the planned concepts phase, a risk verification and judgement are carried out. The purpose of this, is to overcome any mistake or poor assessment over the risks made at the actual analysis phase. Usually, this control of risks is made by a different group than the one involved in the phase before (actual analysis).

A more formal preparation over the risk profile is done in the classification phase. Criteria for classifying and evaluating the risks are created at this phase. The classification is done in terms of the risk types and the evaluation is done in terms of the risk impacts.

Finally, the documentation phases integrate the work done through the other phases in order to create a risk management plan. The outcome of the risk management process shown in figure 3.2 is a plan. This risk management plan is orientated towards the company process and structure. Building this kind of plan demands quite a lot of resources and the decision of choosing the right plan is therefore up to the top managers.

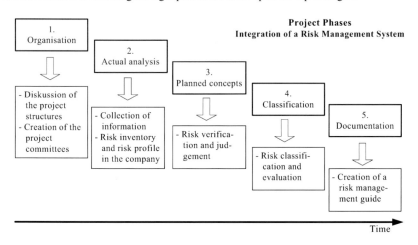

Figure 3.2   Process of a Risk-Management conception (Form and Diederichs 2001)

3.2 Specific risks in construction companies

As with every other kind of industry, the construction industry offers its particular and specific risks arising from wide and diverse scenarios. For example each project takes place in a different location. This different location will give the projects different risk variables which might be related to the human resources available, the transport system available, the availability of space, etc.

In construction companies there are two types of risks: Those related to the company and those related to projects. The main difference between these is that company risks include corporate and strategic risks (top and middle management) whereas project risks, are those which affect individual projects. For example, a typical company risk can be poor productivity of subcontractors. In this case the company has to decide what to do with this risk, either to cancel the contract or to put remedy in some way. A typical project risk can be: delay of setting up the crane. In this case the project manager has the responsibility to accelerate the activities related to the use of the crane, either to increase the number of crews or to delay some activities.

Another situation that might arise in relation to specific risks is for example when two construction companies (of different size) take the same project (in concept). The risks which will arise, will be specific for each company because the capability of each company to deal with the risks is different, this is due to the size. Table 3.2 shows typical project risks for construction companies.

Political Risks	Social Risks	Environmental Risks	Management Risks	Financial Risks
Strikes	Social strikes	Water and air pollution	Organisation structure	Currency fluctuation
Foreign corrupt practices	Society support to project	Epidemics	Bad planning decisions	Credit risk
Legislative changes	Real social benefits	Negative Env. Impacts	Contract disputes	Unstable economy
Tariff policies	Unstable society conditions	Fauna diseases	Bad project management	High inflation
Domestic policies	Low security measures at the construction site	Env. Hazards Regulations	Employee relations towards the company	Bad cash flow management

Table 3.2 Project specific risks

One interpretation of company and project risks is shown in figure 3.3.

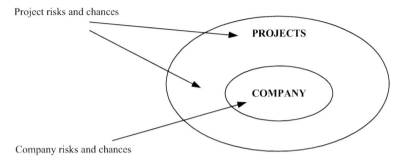

Figure 3.3   Risk Management Environment in the Construction Industry

The reason why the area which represents "Project Risks and Chances" is bigger than the "Company Risks and Chances" (Figure 3.3) area is because construction companies are always project-orientated. In other words, the future of the construction company in relation to its finance, is strongly linked to the success of the projects.

An alternative view of the company is shown in Figure 3.4.

Figure 3.4   Company Environment within Risks and Chances

This figure shows the company interaction with different risks and chances. It introduces the concept of market risks and chances. These are part of the company's risk, but are also part of the possible uncertainty involved for example in construction projects undertaken in unstable countries. Market risks and chances also arise for projects. For example, the project can be affected by market risks where there is an unexpected change in material costs. Chances for both, the company and the project also exist.

The demand of projects, either public or private is related to the market. The market as well is related to the viability of construction companies to tender for the projects in the market.

3.3 Risk management systems in construction companies

Risk management has been considered in construction companies for a considerable time. Nowadays its use is being formalised through the development and creation of well designed processes recommended for use within different industries (for example the AS/NZS 4360)[19].

As in any other business, construction companies have an organisation structure. Within this structure, the objectives of the company are developed and controlled. For that reason, different types of departments (planning, cost estimation) can be identified within the structure. However, to include a risk management department is not an easy task; mainly because not every company has the same interest towards risk and as well the resources to implement one.

Because of the way construction companies exist, their organisation structure is designed for better control of projects. In other words, in construction companies two types of organisation structures exist. One is fully dedicated to organise the company matters and the other one is fully concentrated for the project matters. The decision either to create a new department just for managing risks, or to include the risk management functions in one of the existing departments is a decision which must be made by to the company board of directors.

It is important to keep in mind the difference between a method and a system. Table 3.3 shows the differences between risk management methods and systems.

CONCEPT	DEALS WITH:
Risk Management Methods	Specific project and company risks, concentrates in small or determine areas of a project or company (some projects or company departments). Depending of the method cannot be applied to all types of construction projects (because of the data available). Deals with the projects future outcomes (by forecasting).
Risk Management Systems	The company roles and measures to control risks, integrates two or more company departments, concentrates in all the company organisational structure, is applicable to every type of construction company, works as the company's permanent alert system, deals with the company's future outcomes.

Table 3.3   Risk management methods and systems comparison

---

[19] AS/NZS 4360 (1999). Risk Management.

It can be seen, that risk management methods used in construction companies concentrate on future project performance, behaviour and environment. Risk management methods feed the risk management system with their results. The risk management system then works through the company's structure and management levels to improve the company policies against risks.

Having a risk management department in a construction company is not an easy task. Contractors on many occasions are not willing to pay to establish new structures within their own enterprise. This is more critical, when the contractor does not have any knowledge about risk management.

A more practical way, and perhaps easier and less costly, is to involve the project management team in the risk management tasks. The project management team, are the people who know the project and by following the logic, the project team can contribute more to identifying, analysing and proposing solutions to risks.

One of the requirements of a risk management system in a construction company is, that this system must be interactive (interactive in this case means a dynamic system for controlling the results). A dynamic system offers the advantage of comparing the results and obtaining an output, which is usually the difference between the planned measures and actual measures (control is a fundamental part of risk management because it allows the system to measure its effectiveness on its responses). Figure 3.5 describes an interactive risk management process, under this process, back-propagation (comparison of results) can be done as a good measure to control the quality and the results of the whole system.

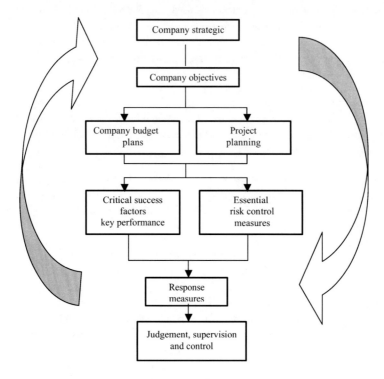

Figure 3.5   Interactive Risk Management Process (PwC)[20]

Bauch[21] (Figure 3.6) describes the risk management process in construction companies, defining the planning phase to be together with the risk identification, analysis and response. Risks which have a relation to the planning phase are cited as the most relevant ones. However, there are some risks (unforeseen risks) that are outside the planning phase of every construction project. For example, earthquakes, floods, hurricanes, political riots, economic crises, etc. Some risks can be envisaged at the planning stage of a project. These include risks which are related to the project cash flow, possible technical problems, problems with the equipment and plant, legal problems, etc. Also, while using contracts (like FIDIC) the contractor can have some protection not only with risks that can be "easily" identified, but also with the ones called "unforeseen risks" or "Acts of God".

---

[20] PwC Deutsche Revision (1999). Finanzwirtschaftliches Risikomanagement deutscher Industrie- und Handelsunternehmen.

[21] Bauch, U. (1994). Beitrag zur Risikobewertung von Bauprozessen.

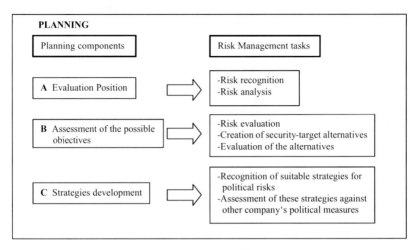

Figure 3.6 Components of the planning phase (Bauch)[22]

3.4 Risk management and organisation

The importance of having a risk management system within the structure of an organisation will be always discussed. The reason is that for some managers (old fashioned) represents perhaps a loss of money and time. Nevertheless, one of the main tasks of the management levels is to decide, if this kind of system is required.

The tasks of the Top-Management, Middle-Management and Operative-Management (see Figure 3.7), represent the dynamic activities of risk management within an enterprise. They interact and receive feed-back from each other as a cycle process through the top and bottom levels of the firm's organisational structure

---

[22] Bauch, U. (1994). Beitrag zur Risikobewertung von Bauprozessen.

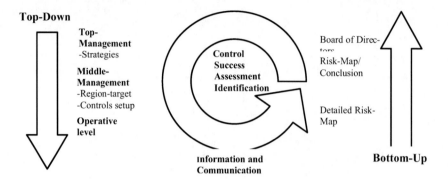

Figure 3.7 Organisation of risk management as a rule-cycle (PwC)[23]

3.5 Company management tasks

The management tasks of a company, represent its operational system. The contents or activities of each task depend on its management level, as follows:

Top-Management:

- long term planning
- company financial cash flows
- long term investments
- overall decision-making
- subjective decisions

Middle-Management:

- company and project control
- middle term planning
- company and project cash flows
- risk management tools

Operative-Management:

- short term planning
- project cash flows
- short investment of resources, personal and equipment
- planning techniques
- total project control

---

[23] PwC Deutsche Revision (1999). Finanzwirtschaftliches Risikomanagement deutscher Industrie- und Handelsunternehmen.

From this example of management tasks can be observed that within the middle management level the use of the risk management tools occurs. That means that it is on this middle management level where, usually, the risks towards the company and its projects are commonly identified, analysed and when is possible, also quantified.

The chart (Figure 3.8) shows the progress that organisations make as they become more sophisticated and expert at managing risk. Initially, risk management activities are ad-hoc and depend on the activities of specific individuals. Efforts may be confined to the identification of risk. Risk management becomes repeatable when there is less dependence on individuals.

The results (Figure 3.8) show, that for almost all organisations, there is a gap between where they currently are and where they would like to be. This demonstrates a commitment to risk management and an understanding that there are gains to be made from deepening and broadening the process. It also shows, that companies have made progress against the Turnbull (British report over risk management, refer to Teji, 2000) requirements to establish a process and policy and dedicate resources to risk management. But now they are up against the hurdles of embedding processes and policies throughout the organisation, adopting consistent risk measures and limits, and devising informative reports (Teji)[24].

---

[24] Teji, T. (2000). Mastering risk beyond Turnbull: Risk Consulting.

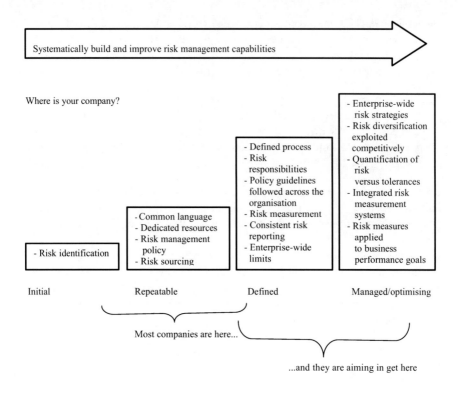

Figure 3.8   Risk Management and the Companies (Teji 2000)

The experience of the United Kingdom regarding the application and use of the risk management theory, describing the benefits of systematic risk management, the risk management process and some points for success can be summarised as follows:

Risk management should help to (CIRIA)[25]:

> ➢ save money
> ➢ reduce accidents
> ➢ deliver projects to time
> ➢ reduce the chances of litigation
> ➢ improve the morale of employees
> ➢ enhance corporate reputation

---

[25] CIRIA (2002). A simple guide to controlling risk.

The benefits offered by the management of risks, are a main strategic goal for company management tasks. They are wide enough to cover different aspects of the company; aspects like financial (save money), safety at work (reduce accidents), marketing (improve the morale of your employees), etc.

The risk management process adopted (CIRIA)[26] is shown in Figure 3.9:

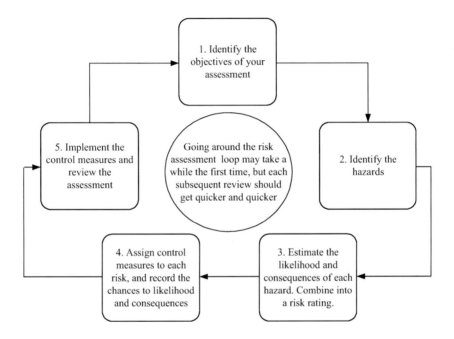

Figure 3.9   Risk Management process

Some points for the success of a systematic risk management are (CIRIA)[26]:

- Risks are unique
- Make your risks explicit
- Focus on the risks that matter
- Apply your judgement
- Keep the process running
- Get the right team
- Know your regularity duties for risk assessment
- Learn step by step

---

[26] CIRIA (2002). A simple guide to controlling risk.

- Focus on success
- Top tips

As a complement to risk management theory, the following figures (figure 3.10) developed by the Joint Standards Australia/ Standards New Zealand Committee OB/7 on Risk Management, described a risk management overview with its risk management process. The document is entitled AS/NZS 4360:1999 Risk Management.

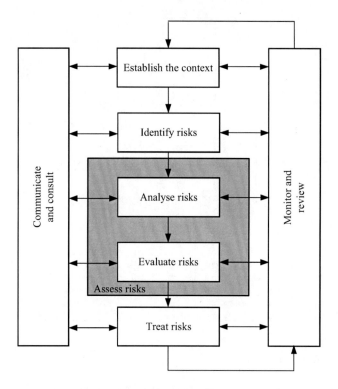

Figure 3.10 Risk Management Overview (AS/NZS 4360)[27]

---

[27] AS/NZS 4360 (1999). Risk Management.

Risks and Chances in Construction Companies 35

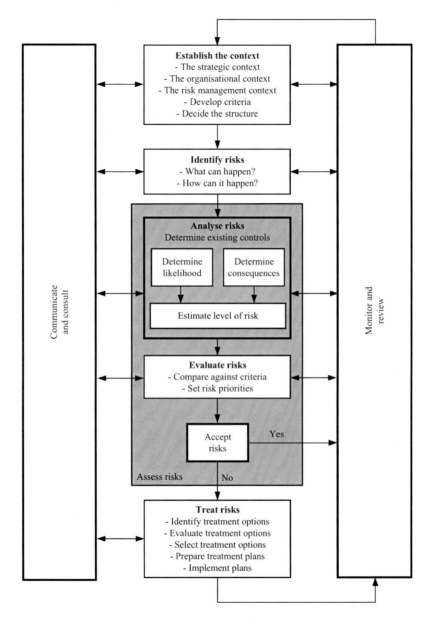

Figure 3.11 Risk Management process (AS/NZS 4360)[28]

---

[28] AS/NZS 4360 (1999). Risk Management.

Although both risk management processes (from CIRIA, see Figure 3.9 and from AS/NZS, see Figure 3.11) vary in concepts, they follow the three main steps of the formal risk management process, which are: identification, analysis and response. Looking at the end for benefits as: risk-cost reduction, more management control over the possible risks, company success, bigger profits, etc.

3. 6   Risk management as a part of the company management

Risk management occurs part in the company environment. In the construction industry, in some cases, risk management acts as controller of the company's decision making because of the relation between construction companies and their projects. In another cases, risk management acts as a consultant because it advises the company in "planning the planning", in other words, risk management provides the company with early advice of the possible outcome of a future project. Forgetting about risk management as a part of the company's vital functions, means forgetting about quantifying the possible costs of the risks and the opportunities.

As an example of the importance of risk management within the company management, the Canadian Risk Management standard (CSA)[29] outlines the risk management decision process as shown in Table 3.4.

The decision process illustrated, is intended to assist decision makers to acquire, analyse and evaluate the information needed to make decisions in areas affected by risk. The process is designed to help decision makers arrive at informed judgements as to the significance of a risk, what level of the risk is deemed acceptable, what level of control might be appropriate, and how to communicate about the risk with stakeholders. Further, it outlines methods of establishing specific actions that may be desirable with respect to the risk, and implementing and checking the effectiveness of those actions (McCallum and Fredericks)[30].

---

[29] CSA (1996). Risk Management: Guideline for Decision-Makers.

[30] McCallum, D., and Fredericks, I. (1997). Linking Risk Management and ISO 14000.

Initiation	- Define the problem or opportunity - Identify Risk Management Team - Assign responsibility, authority and resources - Identify potential stakeholders
Preliminary Analysis	- Define scope of the decision(s) - Begin Stakeholder Analysis - Begin to develop Risk Information Base - Identify possible exposures to loss using risk scenarios
Risk Analysis	- Estimate frequency of risk scenarios - Estimate consequences of risk scenarios - Refine Stakeholder Analysis through consultation - Update Risk Information Base
Risk Evaluation	- Risk Management Team meets to integrate the information from Risk Analysis, including costs - Integrate benefits and update Risk Information Base - Assess acceptability of the risk
Risk Control and Financing	- Identify feasible risk control options - Evaluate risk control options in terms of effectiveness, cost, etc. - Assess stakeholder acceptance of residual risk - Evaluate risk financing options - Assess stakeholder acceptance of proposed action(s)
Action	- Implement chosen control, financing and communication strategies - Risk Management Team evaluates effectiveness of risk management decision process - Establish ongoing monitoring process

Table 3.4   Steps in the CSA-Q850 Risk Management Decision Process, detailed version, from CSA[31]

A risk-systematic approach is provided by the consulting company Arthur Andersen (see figure 3.12). A systematic approach like the Arthur Andersen Business Risk Management Process (BRMP) can assist an organisation to "health check" its processes as well as to thoroughly embed risk management and build up risk management capabilities.

The BRMP identifies seven different elements of effective risk management (Teji)[32]. These are (see Figure 3.12):

- Establish the process
- Assess business risks
- Develop risk management solutions
- Design/implement capabilities

---

[31] CSA (1996). Risk Management: Guideline for Decision-Makers.

[32] Teji, T. (2000). Mastering risk beyond Turnbull: Risk Consulting.

- Monitor performance
- Continuously improve capabilities
- Information for decision making

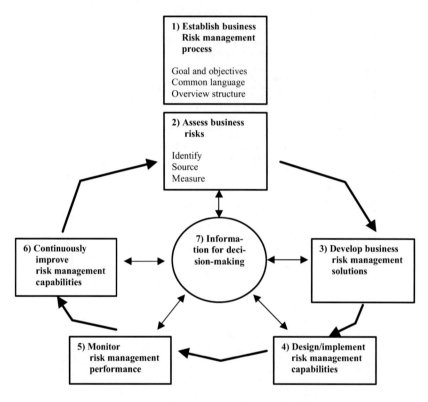

Figure 3. 12   Business Risk Management Process (BRMP) (Teji)[33]

3.7   Systematic solutions

Establishing a system for controlling the risks and chances in a company is the best way to deal with the uncertainty of the project. However, it is clear that doing this is not an easy task because each construction project is unique. In addition, the task gets more difficult when the enterprise undertakes different types of projects. Another problem is located within the organisational structure of the company. Normally within a construction company it is possible to find an area or department that is in charge of the manage-

---

[33] Teji, T. (2000). Mastering risk beyond Turnbull: Risk Consulting.

ment of the project, under this branch, a variety of activities are done such as project evaluation, feasibility studies, facility management, quality management, planning and control, project finance and risk analysis. The last activity mentioned, is carried out in most of the cases without a systematic approach or process. Sometimes the analysis of risks is done with the help of probabilistic methods but in most cases only by subjective decisions.

A formulation made from Mawdesley[34], establishes that each attempt to deal with risks and uncertainties must posses certain basic procedures, like:

- identification of the uncertain aspects of the project
- estimate the meaning of each particular aspect, regarding its effects and probability of occurrence
- to work out the balance between chances and risks
- acceptable control measures over the uncertain aspects of the project
- application of measures, in order to be able to reduce the individual uncertainties

A systematic approach of risk management is vital for obtaining a balance between risks and chances. Development a defined strategy of how to tackle the risks and the chances, will safeguard the company's financial stability.

---

[34] Mawdesley, M., Askew, W., and O'Reilly, M. (1997). Planning and controlling construction projects: The Best Laid Plans.

4 Risk Management Methods in the Construction Industry

This chapter contains a short description and discussion of the different risk management methods available for the evaluation of risks in construction projects.

4.1 Net Present Value-at-Risk Method in Infrastructure Project Investment Evaluation (Sudong Ye and Robert Tiong)[35]

4.1.1 Introduction, description and discussion

Strategic capital investment decisions are crucial to a company. The decision to invest in privately financed infrastructure projects requires careful consideration, because they are exposed to high levels of financial, political, and market risks. The project appraisal methods should incorporate analysis of these risks. A number of capital-investment decision methods can take risks into account, but each of them focuses on different factors and has its limitations. Thus, a more rigorous method is needed. A systematic classification of existing evaluation methods shows, that it is possible to develop a new method, the Net-Present Value-at-Risk (NPV-at-risk) method by combining the weighted average costs of capital and dual risk-return methods (Sudong Ye and Robert Tiong)[35].

First, it is important to explain what the NPV (Net Present Value) technique is about. The NPV method compares the cash inflows and outflows by restating them to values as of a common date, most logically the present date/time. The decision rule is to accept a project if its NPV is positive (or its ratio of PV of cash inflows to the investment is higher than the alternatives). In other words, the NPV method is a way of comparing the value of money now with the value of money in the future.

The method of NPV-at-risk is defined as a particular NPV that is generated from a project at some specific confidence level, that is, the minimum expected NPV with the given confidence level. In other words, NPV-at-risk is the value at which a % of possible NPVs are smaller and 1- a % are larger. According to the definition of NPV-at-risk, the following decision rules can be derived: "the project is acceptable with a confidence level of 1- a if NPV-at-risk at the given confidence is greater than zero; otherwise, it is unacceptable". Alternatively, the project is acceptable if the computed confidence level at the point zero NPV is equal to or greater than the predetermined confidence level; otherwise, it is unacceptable. The NPV-at-risk method aims to calculate the value that the project's NPVs will be greater than, with the probability corresponding to the given

---

[35] Sudong, Y., and Tiong, L.K. R. (2000). NPV-at-risk method in infrastructure project investment evaluation.

confidence level. It involves the determination of discount rate and the generation of cumulative distribution of possible NPVs (Sudong Ye and Robert Tiong).[36]

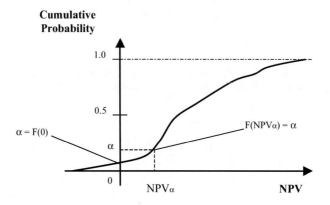

Figure 4.1   Calculation of NPV-at-Risk and Confidence Level Based on Cumulative Distribution Function (Sudong Ye and Robert Tiong)[36]

The NPV-at-risk method produces a single NPV value from a range of outcomes at a given confidence level (Figure 4.1). The NPV-at-risk method takes the following factors into account: (1) All the possible returns resulting from uncertainty; (2) the time value of money; (3) the impact of financing methods; and (4) various risks associated with BOT (Build-Operate and Transfer) projects (Sudong Ye and Robert Tiong)[36].

4.1.2   Conclusion

The combination of methods used to create the NPV-at-risk method is in overall terms a very good alternative for the evaluation of infrastructure projects this is because NPV-at-risk incorporates both, risk and financing methods. However, the usefulness of the results depends entirely of the simulation process. This method can provide good assistance for the decision making for infrastructures projects that are undertaken in countries which have economic problems or where the political and social risks are expected to play a relevant role within the construction of the project. In general terms, NPV-at-risk is a probabilistic method that can be a good alternative to evaluate as well the

---

[36] Sudong, Y., and Tiong, L.K. R. (2000). NPV-at-risk method in infrastructure project investment evaluation.

money in the life cycle of a project, because it takes into account the money concept (capital investment) throughout the duration of the project.

The question of what an "infrastructure project" has to do with a construction company is clarified by the following statement. The term "infrastructure project" was used only to show the possible results that can be obtained with the NPV-at-risk method. The method can be applied to any type of project.

A critical point of using NPV-at-risk in construction projects in developing countries is the discount rate. Under these situations, considering a stable or constant discount rate for some period of time is a matter of discussion. The risks involved in choosing the discount rate and its changes over the life of a project are considerable. These must be taken into account in the simulation.

4.2 Value at Risk (Linsmeier and Pearson)[37]

4.2.1 Introduction

Value at risk is a single summary and a statistical measure of possible portfolio losses. The question of what a portfolio has to do with a construction company is answered. For a company, in this case a construction company, it is very important to develop a financial portfolio. With this, it is possible to know the financial behaviour of the company at any future time. When considering capital investments, talking about "portfolio losses" it is as well very important for the company because the possible financial risks throughout a future period of time can be identified and analysed. A considerable part of these risks will come from the construction projects. This type of analysis is encouraged when a contractor is planning to take works in developing countries.

Losses greater than the value at risk are suffered only with a specific small probability. Subject to the simplifying assumptions used in its calculation, value at risk aggregates all of the risks in a portfolio into a single number suitable for use in the boardroom, reporting to regulators, or disclosure in an annual report. Once one crosses the hurdle of using a statistical measure, the concept of value at risk is straightforward to understand. It is simply a way to describe the magnitude of the likely losses on the portfolio.

---

[37] Linsmeier, J. T., and Pearson, D. N. (1996). Risk Measurement: An Introduction to Value at Risk.

4.2.2 Value at Risk methodologies

Historical Simulation: Historical simulation is essentially a simple, theoretical approach that requires relatively few assumptions about the statistical distributions of the underlying market factors. The procedure is illustrated with a simple portfolio consisting of a single instrument, the 3-month forward contract for which the distribution of hypothetical Mark-to-Market profits and losses is shown in Figure 4.2. In essence, the approach involves using historical changes in market rates and prices to construct a distribution of potential future portfolio profits and losses (Figure 4.2), and then reading off the value at risk as the loss is exceeded only 5% of the time (Linsmeier and Pearson).

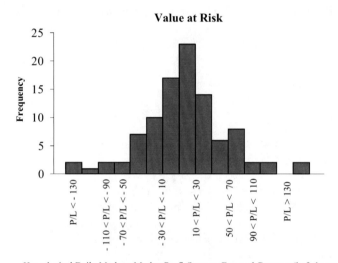

Hypothetical Daily Mark-to-Market Profit/Loss on Forward Contract (in $ thousands)

Figure 4.2 Histogram of Hypothetical Daily Mark-to-Market Profits and Losses on a Forward Contract (Linsmeier and Pearson)

Variance-covariance approach: The variance/covariance approach is based on the assumption that the underlying market factors have a multivariate Normal distribution. Using this assumptions (and other assumptions detailed below), it is possible to determine the distribution of Mark-to-Market portfolio profits and losses, which is also normal. Once the distribution of possible portfolio profits and losses has been obtained, standard mathematical properties of the normal distribution are used to determine the loss that will be equalled or exceeded x percent of the time, i.e. the value at risk (Linsmeier and Pearson).

	Historical Simulation	Variance/Covariance	Monte Carlo Simulation
Able to capture the risks of portfolios which include options?	Yes, regardless of the options content of the portfolio	No, except when computed using a short holding period for portfolios with limited or moderate options content	Yes, regardless of the options content of the portfolio
Easy to implement?	Yes, for portfolios for which data on the past values of the market factors are available	Yes, for portfolios restricted to instruments and currencies covered by available "off-the-shelf" software. Otherwise reasonably easy to moderately difficult to implement, depending upon the complexity of the instruments and availability of data.	Yes, for portfolios restricted to instruments and currencies covered by available "off-the-shelf" software. Otherwise moderately to extremely difficult to implement.
Computations performed quickly?	Yes	Yes	No, except for relatively small portfolios
Easy to explain to senior management?	Yes	No	No
Produces misleading value at risk estimates when recent past is atypical?	Yes	Yes, except that alternative correlations/standard deviations may be used.	Yes, except that alternative estimates of parameters may be used.
Easy to perform "what-if" analyses to examine effect of alternative assumptions?	No	Easy able to examine alternative assumptions about correlations/ standard deviations. Unable to examine alternative assumptions about the distribution of the market factors, i.e. Distributions other than the Normal.	Yes

Table 4.1   Comparison of Value at Risk Methodologies (Linsmeier and Pearson)

Monte Carlo Simulation: The Monte Carlo simulation methodology has a number of similarities to historical simulation. The difference of the Monte Carlo simulation with the historical simulation is that, instead of choosing information generated by observing changes in the market along some periods of time; all this, in order to create a model for representing profits and losses. The Monte Carlo simulation chooses a statistical distribution for assuming the knowledge of the behaviour of these changes in the market as an approximation. Then, a pseudo-random number generator is used to generate thousands or perhaps tens of thousands of hypothetical changes in the market factors. These are then used to construct thousands of hypothetical portfolio profits and losses on the current portfolio, and the distribution of possible profit or loss. Finally, the value at risk is then determined from this distribution (Linsmeier and Pearson).

Table 4.1 provides a guide for the future user of Value-at-risk with the Historical simulation, Variance-Covariance method and the Monte Carlo simulation.

4.2.3 Conclusion

This technique is useful within the company's risk level (market risks), but not in the case of the project's risks. As well, this technique might be of help for international contractors who mainly operate their business units in unstable economies, where part of the strongest impact of the uncertainty deals with foreign currency prices and interest rates (market movements).

Value at Risk is a risk management technique, which can be used for example while developing a "prompt list" for identifying the most probable risks (at company level). Further on, using these risks together with the Monte Carlo simulation in order to identified the variables which offer the most critical risk impacts to the company.

A considerable disadvantage of Value at Risk is, that its application demands a high knowledge of financial and market terms. This situation can represents a problem to the project or construction manager who is mainly aware about other technical and construction related factors, which are also important and relevant for the company.

Value at Risk until now has been used only within the field of financial risks. The question is, if this technique is flexible enough to be used in other areas of risk, such as, contract risks, technical risks and management risks.

Another point to discuss, is that Value at Risk does not offer a single number (risk value) as an answer regarding the risk cost. Value at Risk provides some frequencies (see Figure 4.2) and, in each case the (Mark-to-Market Profit/Loss) is evaluated towards the probable value of the risk of each case.

4.3 Genesis of Management Confidence Technique (Jaafari)[38]

4.3.1 Introduction

Management Confidence Technique (MCT) is a technique for assessing the risk of project failure. It has been developed as a practical tool for formulation of early strategies to mitigate the effects of perceived constraints affecting any project in question. Judicious use of MCT will enable the promoter of a project or his appointed manager to consider all potential options available for dealing with a given constraint and, perhaps more importantly, to consider the risk resulting from the combined effects of all perceived constraints upon his or her project (Jaafari)[38].

One of the features of MCT is its ability to temper, up front, the manager's approach to a new and potentially risky project. However, the application of MCT is not limited to the formulation stage of projects; it can be used at any stage to assess the project's propensity to succeed or fail, given the perceived effects of the remaining constraints (Jaafari)[38].

4.3.2 Objective and requirements

The chief objective is to devise a technique, to be known as the Management Confidence Technique, MCT, which enables and overall evaluation to be made of the probability of a project failing to meet its original objectives because of the combined effects of all its likely constraints. The MCT requirements are as follow (Jaafari)[38]:

- ➤ The proposed management technique must aid the planner/manager in his or her effort to identify all the constraints that may affect a project,
- ➤ Any framework adopted should enable the relative effect of various constraints to be taken into consideration,
- ➤ It must also allow the combined effects of the relevant constraints to be evaluated,
- ➤ Although not quantitative the technique, must nevertheless, lead to a tangible measure, say an estimate of the probability of overall failure, for a set of assumed strategies, to give the manager a picture of the project's likely outcome and to try out various strategies in search of an optimum solution,
- ➤ Although evaluation of the effect of the relevant constraints may be objective, the technique should yield consistent results in repeated cycles of evaluation, provided it is applied prudently and assuming that the user has sufficient experience in judicious use of the technique,

---

[38] Jaafari, M. (1987). Genesis of management confidence technique.

➢ Force majeure events, which by definition are beyond contemplation and control of managers, must be excluded,

The MCT's authors have developed and included three types of different constraints to be evaluated, while applying this technique. These constraints are: (1) Project-related, such as project size, ground conditions, process technology; (2) management-related, such as poor control system, lack of leadership, inappropriate organisational structure; and (3) environment-related, such as socio-economic variables plus legal and political, fiscal, and educational constraints. The following table is used to score the constraints.

Score	Description
0	No effect
1	Perceptible effect
2	Significant intensity
3	Fair intensity of effect
4	Near medium intensity
5	Medium intensity of effect
6	Above medium intensity
7	Near high intensity
8	High intensity
9	Near extremely high intensity
10	Presence or intensity of effects will be extreme

Table 4.2   Scale of Score for each Constraint (Jaafari)

The formula to calculate the weight of each type of constraint is as indicated in Formula 4.1. For this case, this formula is for the constraints classified as "project related". For calculating the overall constraints value, Formula 4.2 is used:

$$(4.1) \quad V_1 = \frac{m \left(\sum_{i=j}^{m} W_i S_i \right)}{\sum_{i=j}^{m} 10 W_i}$$

$$(4.2) \quad V = V_1 + V_2 + V_3$$

Where:

V_1 = combined weighted and adjusted subjective value representing all constraints classed as "project related"

m	=	total number of project-related constraints
W_i	=	the perceived weight for constraint i
S_i	=	score awarded to constraint i
V_2	=	combined weighted and adjusted subjective value representing all constraints classed as "management related"
V_3	=	combined weighted and adjusted subjective value representing all constraints classed as "environmental-related"

The number 10 used in Formula 4.1, represents the total number of different scores used for each constraint. When the total value of the project's constraints weight (V) is obtained, then by using a scale factor ß [%] the probability of the project overall failure is calculated. This is done by finding the V value on Figure 4.3 and then relating it to the "Overall Failure Risk ß [%]" value.

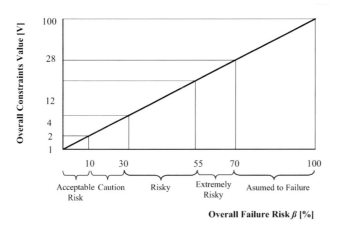

Figure 4.3   Relationship between Overall Constraints Value (V) and Overall Failure Risk (ß) (Jaafari)

Examples of the types of constraints that can be considered in this technique, are shown in tables 4.3, 4.4 and 4.5.

Type		Actual constraint
Site and locality		
	P_1	Topography and natural barriers within the site area
	P_2	Geology, nature of site materials, groundwater, latent conditions
	P_3	Climate and severity of the elements of nature, weather-windows, etc.
	P_4	Production space, construction access, maneuverability of standard equipment
	P_5	Considerations regarding environmental impacts (ecology, heritage, noise, fume, and run-off control, disruption to local way of life, business, and traffic, etc.
Nature of works		
	P_6	Size of works, it's spread around the site area
	P_7	Nature (type and complexity) of works
	P_8	Stringent quality assurance/control requirements
	P_9	Innovative construction methods

Table 4.3   Suggested Initial List of Project-Related Constraints (Jaafari)

Type		Actual constraint
Contractual and legal		
	M_1	Harsch (autocratic) contract conditions
	M_2	Adversarial elements, lack of an appropriate mediation/arbitration provision
	M_3	Lack of performance motivators
	M_4	Divergence of contractual terms from project objectives
	M_5	Complicated contractual terms, matters, procedures, etc., occupying inordinate amount of management time
	M_6	Complicated joint venture structure
Human factors		
	M_7	Lack of effective leadership
	M_8	Absence of trust and mutual respect
	M_9	Absence of group culture and standards
	M_{10}	Unrealistic expectations from subordinates
	M_{11}	Inflexible attitudes to the need for changes and adaptation

Table 4.4   Suggested Initial List of Management-Related Constraints (Jaafari)

Risk Management Methods in the Construction Industry 51

Type	Actual constraint
Economic commerce	
E_1	General business system and the structure of the economy, e.g., private sector's functions, extent of competitiveness in the market, foreign exchange control, and government's control of the economy by direct and indirect intervention
E_2	Capital and credit availability and control of the monetary inflation
E_3	"Factor-endowment," which is indicative of the extent the country benefits from natural resources and availability of adequate skilled and semiskilled labor
E_4	The market size (for long-term investment commitment)
E_5	The existence of local industries. The extent of their development, and their cost-effectiveness. Also their potential for further development
E_6	The local average productivity level
Sociological-cultural	
E_7	The general values choose by the society, e. g., is material wealth considered the main source of satisfaction or is living considered as means of preparation for life after dead

Table 4.5   Suggested Initial List of Environment-Related Constraints (Jaafari)

4.3.3 Conclusion

The Management Confidence Technique is a very good attempt to integrate three sub-areas of risk management into one. In other words, to integrate the value of the risk of these areas into a single number/ classification of how risky a project is.

As a management tool, this technique is quite attractive. However, there is no example provided where the cost of the risk is appraised in terms of money. It is important to know how risky a specific project is or might be, depending on the constraints used with Management Confidence Technique (Figures 4.3, 4.4 and 4.5), but it is more relevant to the contractor to know the cost of the risk.

With the Management Confidence Technique it is possible to know the Overall Failure Risk (ß in %) of the project. This number advises the contractor in deciding if it is a good idea to undertake a project. In other words, ß classifies the risk in the project. Perhaps it is possible to link the ß value to money terms, in this case this technique might provide more concrete answers about risk. The use of other constraints is encouraged (like Contract Constraints, Client Constraints, Guarantee Constraints, Subcontractors Constraints, etc.) in order to give reliability to the method. In other words, to increase the range of constraints will give the technique more possibility to be applied in with different type of projects.

The structured approach of the Management Confidence Technique towards the qualitative evaluation of the three main types of constraints is beneficial, because it encloses different types of factors that can affect a project. The structure is also flexible and the addition of new types of constraints is possible.

A disadvantage of the method is, that it evaluates the risk impact of the project in the qualitative manner. It provides an answer which is related to the impact of the "Overall Failure Risk" of the project, but this value is not related to the project costs or to the probable risk costs.

The technique as well needs further application in real life; in other words, the techniques needs to be used within real construction projects in order to test the usefulness of the method; and based on this, to decide the reliability of Management Confidence Technique as a risk management tool.

However at the research stage, Management Confidence Technique provides a good approach to identify and evaluate different types of risks and to obtain a single value of how risky the project can be based on the constraints used. As an overall commentary over the use of this technique, it can be said that the results obtained are quite useful especially when the project concerned represents a huge investment of capital or when the project is undertaken in developing countries. However, subjective probability is used as a "help tool" to determine the weight and scores of the constraints. The author's idea is supported in relation to the use and promotion of this approach on new projects, although as a recommendation, this approach might be appropriate for top management levels.

4.4 Estimating using risk analysis (ERA) (Mak and Picken)[39]

4.4.1 Introduction

A contingency is an amount of money used to provide for uncertainties associated with a construction project. Traditionally, it is a percentage addition on top of the base estimate. Estimating using risk analysis (ERA) is a methodology that can be used to substantiate the contingency by identifying uncertainties and estimating their financial implications.

---

[39] Mak, S., and Picken, D. (2000). Using risk analysis to determine construction project contingencies.

In an attempt to deal with the determination of contingencies in a more analytical way the Hong Kong Government implemented a technique called Estimating using Risk Analysis (ERA) in 1993 (Mak and Picken).

ERA is used to estimate the contingency of a project by identifying and costing risk events associated with a project. The starting point for the ERA process is a base estimate, which is an estimate of the known scope and is risk free. The contingencies as determined by the ERA process are added to the base estimate. The first step in the ERA process is to identify risks. There risk items are then categorised as either (1) fixed; or (2) variable. For each risk event, an average risk allowance and a maximum risk allowance are calculated. The relationship between risk category and risk allowance is shown in table 4.6 (Mak and Picken).

Type of risk	Average risk allowance	Maximum risk allowance
Fixed risk	Probability x maximum cost	Maximum cost
Variable risk	Estimated separately	Estimated separately
Assumption	50 % chance of being exceeded	10 % chance of being exceeded

Table 4.6   Relationship between Risk Allowance and Risk Category ERA (Mak and Picken)

Figure 4.4 shows a typical ERA worksheet, once that all the risk events were identified and their average and maximum risk allowance calculated.

ERA Calculation						
Project: Construction of the Central Library					Date: 2 March 1995	
Client: Urban Council					ERA Run: 1	

Risk	Type	Probability (Fixed Risks Only)	Average Risk Allowance $	Max. Risk Allowance $	Spread (5) - (4) $ M	Spread Squared $ M
Design Development	V		8,400,000.00	12,600,000.00	4.20	17.64
Additional Space	F	0.70	11,760,000.00	16,800,000.00	5.04	25.40
Site Conditions	V		525,000.00	1,000,000.00	0.47	0.22
Market Conditions	V		4,000,000.00	8,500,000.00	4.50	20.25
A/C Cooling Source	V		250,000.00	1,250,000.00	1.00	1.00
Access Road	F	0.50	250,000.00	500,000.00	0.25	0.06
Addtional Client Requirements	V		1,680,000.00	4,200,000.00	2.52	6.35
Contract Variations	V		8,400,000.00	12,600,000.00	4.20	17.64
Project Coordination	V		500,000.00	1,500,000.00	1.00	1.00
Contract Period	F	0.60	1,000,000.00	1,750,000.00	0.75	0.56
			36,765,000.00			90.13
					Sq. Root	9.49
			Maximum Likely Addition =			$ 9,490,000.00

Base Estimate	= $ 168,000,000.00
Average Risk Estimate	= Base Estimate + Total Average Risk Allowance
	= $ 168,000,000.00 + $ 36,765,000.00
	= $ 204,765,000.00 (21.88 % on base)
Maximum Likely Estimate	= Base Estimate + Average Risk Allowance + Maximum Likely Addition = $ 168,000,000.00 + $ 36,765,000.00 + $ 9,490,000.00
	= $ 214,259,000.00 (27.54 % on Base)

Figure 4.4   Example of ERA Worksheet at Sketch Design Stage (Mak and Picken)

4.4.2 Advantages of ERA

It should be noted, that the use of ERA does not necessarily reduce the total cost of a project. It makes it less uncertain. A distinct advantage of ERA lies in its ability to retain the traditional method of presenting a project cost estimate in the form of a base estimate plus a contingency. It imposes a discipline from the outset to systematically identify, estimate cost, and consider the likely significance of any risks associated with a project. It also aids financial control in having risk and uncertainty costs identified before action is taken to determine precise requirements. Furthermore, the itemised and substantiated contingency forms the basis on which to evaluate the impact of risk and uncertainties upon completion of the project, thereby providing useful data for use in investment appraisal. In short, it is a mechanism for accountability for public money. Rigorously done, it will reduce the usually conservative and excessive percentage add-on contingency and lead to a better allocation of resources (Mak and Picken).

4.4.3 Conclusion

The results of the ERA method when it is applied throughout Public Government Offices are until now acceptable. Although it is necessary to test this method within different projects. As well, there is the possibility to promote and implement the ERA technique with other types of organisations, like for example the World Bank.

In the ERA process, there is a systematic approach; but the way that the "Average Risk Allowance" and the "Max. Risk Allowance" are determined is still in the traditional form. Perhaps with the use of "historical records" this can be improved. ERA is a good technique to use at the middle and low management levels, because many of the risks at these levels can be evaluated.

It is not clear how the method defines that a risk is a "Fixed risk" or a "Variable risk". This is a disadvantage which can cause confusion at the stage of identifying and then classifying the risks. The confusion can provoke that for one project manager a risk can be a "Fixed risk" while for another project manager the same risk can be a "Variable risk". This confusion in cost terms is also considerable.

4.5 Risk and Opportunity Analysis Device (ROAD) (Link)[40]

4.5.1 Introduction

ROAD is a risk evaluation model for building processes, which enables a faster and more efficient analysis of project risks. The risk analysis made by ROAD, is intended to be applied at the supply or calculation phase of the project, helping in detecting the possible scenarios of the project development and the effects of the different entry parameters over the total risk and the setting of appropriate safeguard measures (Link).

---

[40] Link, D. (1999). Risikobewertung von Bauprozessen.

In ROAD, the project is divided into its individual phases and standard and special risks are detected in those project phases. There are two types of period-risks managed in ROAD, active and passive. The active risk period covers the operational phase of a project; in other words, these risks are the ones present during the construction period. The passive risks are presented in a period of the project which belongs to the post-construction phase, usually are risks which appear once the project has been handled to the client or is operated by the contractor.

The starting point of the ROAD model is the specification, the target of ROAD is not to approximate exactly to the real system, but to obtain a high correlation between the forecast outputs and the real outputs. The next procedure is to carry out a simulation, using the Monte Carlo method. The simulation is done with the help of the Crystal Ball programme (see Figure 4.5).

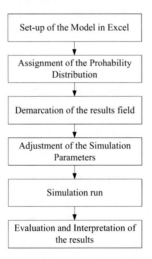

Figure 4.5   Risk-Assessment outflow in Crystal Ball (Link)

4.5.2 Project phases

First, the selection of the project-phases is done. The typical project phases in ROAD are: Feasibility studies, design-planning, submission, advertisement, tendering, supply-check, execution, cost estimating and the operation. After this, an allocation of the project costs to each individual project phase takes place. By doing this, it is easy to know which phase possesses the higher risk cost (see Figure 4.6), depending on the price.

Project Name	Duration	Successor Interception	Successor Interception	Price [Euro]
Total Projects				
Phase 1				
Phase 2				

Figure 4.6   Arrangement of the project-data (Link)

The risk check-list especially developed for ROAD forms the basis of the risk classification. The risks in the check-list can be divided in three categories:

- Standard risks: occur in all building projects with the same probability and similar effects,
- Section risks: occur only in specific sub-ranges,
- Project-specific risks: occur in all building projects with different probabilities and/or different effects,

A classification for measuring the risk-effects in probability terms was also developed, this is shown in Table 4.7:

Is a defined event after the judgement of the Risks especial lists	Then is the subjective probability in %
complete impossible	0 %
extraordinary improbable	1 - 10 %
very improbable	5 - 20 %
fair improbable	10 - 30 %
Improbable	20 - 40 %
constantly improbable	30 - 50 %
constanty probable	40 - 60 %
absolutely probable	50 - 70 %
Probable	60 - 80 %
fair probable	70 - 90 %
very probable	80 - 95 %
extraordinary probable	90 - 99 %
complete probable	100 %

Table 4.7   Weighting of probabilities of subjective estimations related to verbal expressions (Link)

Table 4.7, shows the risk assessment scale of ROAD, based on subjective probability. There are some disagreements in the scale, for example in the event "very probable" the scale is 80 – 95 %, and in the event "extraordinary probable", the scale is 90 – 99 %.

No.	Type of Distribution	Parameters
1	Rectangle	Minimum Maximum
2	Normal	Middle value Standard deviation
3	Triangle	Minimum Maximum Probability value
4	Binomial	Searched quantity Value probability
5	Poisson	Rate
6	Geometrical	Probability
7	Hypergeometrical	Population size Experiment Initial probability distribution
8	Exponential	Rate
9	General	Single value serie Discrete values regions Constant values regions
9.1	Event	Values Probability

Table 4.8   Possible probability distributions of ROAD

The different types of distributions that can be used with ROAD, together with their parameters are shown in Table 4.8. The overall project risk is determined by overlaying the risk phases; this is done by an average value of the total risk distributions. This value is transformed into monetary units by an approximate percentage value, for each phase.

4.5.3   Conclusion

The way used to apply the Monte Carlo and the Latin-Hypercube simulation methods is an alternative approach for the analysis of risks. However, it must be mentioned that especially the use of the Monte Carlo simulation in analysing the risks in construction projects, is not new and again the use of subjective probability is present.

Although each parameter of ROAD was described in detail, it is difficult to find a simple example of a real project. Instead, a hypothetical project was used but the numbers of the results are too wide and this causes or will cause confusion for the project manager (or project evaluator). However, the use of the Crystal Ball program together with the Monte Carlo and Hypercube simulation methods, as well with the risk framework developed by Link is a considerable contribution to the risk management paradigm within construction projects. It is an attempt to quantify the risks in terms of money.

A considerable disadvantage of ROAD is that it has not yet been tested with real projects. This lack of real application provides ROAD with a low reliability and perhaps no usefulness in the construction industry. Another disadvantage is that the example provided by Link has no concrete results of how much the contract or project price was incremented or must be incremented as recommended by ROAD. In addition, the theoretical project example used is not useful for discussing the results in detail.

ROAD proposes to use two types of risks; active and passive. The active risks, are the ones which will appear once the execution of the project start. The passive risks are the ones which are identified at the planning phase. However, in practice, a contractor does not have the enough time to make any kind of analysis once the construction works has started. The reality of any construction project is that once it has been started, it is very difficult to undertake any extra analysis; in fact, it seems very difficult to do risk analysis towards the possible active risks during the planning stage of a project.

The theoretical basis of ROAD is quite complete. However, it seems to be difficult or quite demanding for a contractor, to implement this model within the company operations tasks. ROAD is not only about identifying risks, it is also about learning different parameters of probability which a project manager usually (at the practical work as cost estimator) is not so familiarised. What would be an advantage for ROAD, is perhaps to concentrate on a single model (less range of probabilities, see Table 4.8) and to propose this case for use within contractors.

A knowledge of several types of probability distributions is needed to operate ROAD properly.

4.6 Evaluation of the risk management methods

The intention of Table 4.9 is to show a resume of the versatility and reliability of the risk management methods described in this chapter. As can be seen, no method offers a single number or result about the possible costs of the uncertainty in construction projects. This is the main reason of this research work (see chapters 8 and 9).

60 Risk Management Methods in the Construction Industry

METHOD	Advantages	Disadvantages	Application	Results offered	Practical experience
NPV-AT-RISK (see chapter 4.1)	• Easy to manage • Clear results • Relative no expensive to be implemented • Flexible • Good background	• Needs a good simulation process • Needs more application in different projects • Requires good analytical skills	• Until now, only with infrastructure projects	• Graphic results • Results forecast the possible project profit	• Only with infrastructure projects • Need further promotion in other projects
Value at Risk (see chapter 4.2)	• Can be use within different kind of projects • Relative easy to understand the results • Good for the company financial forecasting	• Requires good knowledge of financial concepts • Needs more application with construction projects • Requires a considerable knowledge of analytical skills	• Until now, only with financial companies	• Graphic results • Results in a single number	• Only with financial markets • Needs further promotion in construction companies
Management Confidence Technique (see chapter 4.3)	• Measures how risky is a project • The technique is well structured • Easy to understand • Good method for middle and top management use	• Takes considerable time to apply • Needs practical use • Relative high subjective method	• Only with empirical construction projects	• Measures the risk in projects but not in money terms	• Needs further use with real projects • Only with empirical projects
Estimating Using Risk Analysis (see chapter 4.4)	• Quantify the possible risk cost in projects as a "Risk Premium" • Relative easy to understand • Good background • Flexible • Good organisation structure	• Highly subjective orientation • Posses no international reputation • Needs to be use in the private sectors	• Only with public projects	• Results in money terms	• Experience with public projects • Needs to be applied within private projects
Risk and Opportunity Analysis Device (see chapter 4.5)	• Offer results in money terms • Use reliable simulation methods • Applicable to different types of projects	• Expensive to implemented • Require high knowledge of simulation processes • Possess no clear building process • Offers no single results • Highly subjective orientation • Requires high analytical skills	• Until now only in one empirical project	• Results in money terms for each phase	• Only in theoretical projects • Needs to be implemented with real projects

Table 4.9   Evaluation of the risk management methods

5 Artificial Neural Networks

5.1 What is an artificial neural network?

Also referred to as connectionist architectures, parallel distributed processing, and neuromorphic systems, an artificial neural network (ANN) is an information processing paradigm inspired by the way the densely interconnected, parallel structure of the mammalian brain processes information. Artificial neural networks are collections of mathematical models that emulate some of the observed properties of biological nervous systems and draw on the analogies of adaptive biological learning. The key element of the ANN paradigm is the novel structure of the information processing system. It is composed of a large number of highly interconnected processing elements that are analogous to neurons and are tied together with weighted connections that are analogous to synapses (Battelle Memorial Institute)[41].

Artificial neural networks offer an approach to computation that differs greatly from conventional methods. Artificial neural networks attempt to use some organizational principles that are used in the human brain (i.e. parallelism, adaptivity, inherent, etc.).

Learning in biological systems involves adjustments to the synaptic connections that exist between the neurons. This is true of ANNs as well. Learning typically occurs by example through training, or exposure to a set of input/output data where the training algorithm iteratively adjusts the connection weights (synapses). These connection weights store the knowledge necessary to solve specific problems. Although ANNs have been around since the late 1950's, it was not until the mid-1980's that algorithms became sophisticated enough for general applications. Today ANNs are being applied to an increasing number of real-world problems of considerable complexity (for example: in quality control, financial forecasting, economic forecasting, process modelling and management). They are good pattern recognition engines and robust classifiers, with the ability to generalize in making decisions about imprecise input data. They offer ideal solutions to a variety of classification problems such as speech, character and signal recognition, as well as functional prediction and system modelling where the physical processes are not understood or are highly complex. ANNs may also be applied to control problems, where the input variables are measurements used to drive an output actuator, and the network learns the control function. The advantage of ANNs lies in their resilience against distortions in the input data and their capability of learning. They are often good at solving problems that are too complex for conventional technologies (e.g., problems that do not have an algorithmic solution or for which an algorithmic solution

---

[41] Battelle Memorial Institute, 1997. Artificial Neural Networks.

is too complex to be found) and are often well suited to problems that people are good at solving, but for which traditional methods are not (Battelle Memorial Institute)[42].

5.2 The structure of the artificial neural network

The three different structures of a generic ANN are (Maren)[43]:

- micro structure: it determines the activity of a single neuron,
- meso structure: it creates the physical organisation of the neurons,
- macro structure: it is a union of different meso structures to solve complicated problems,

5.2.1 Micro structure

The ANN micro structure is the smallest component of the network, that is the j_{th} neuron (see Figure 5.1).

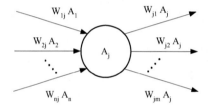

Figure 5.1   Micro structure of ANN (Himanen et al)[44]

The symbols in Figure 5.1 are defined as follows:

- W_{ij} (i = 1, ... n; j = 1, ... m) are the weights of synaptic connections between the j_{th} neuron and the neurons linked to it.
- A_j (j = 1, ... n) are the activations of neurons. Each neuron has its own "activation state" determined by the activity of the neurons linked to it, based on the relationship shown in equation 5.1.
- the formula (5.1) denotes the "activation" function (f) (the so-called transfer function) which allows the transmission of the signals only when the activation of a neuron exceeds a particular threshold value.

---

[42] Battelle Memorial Institute, 1997. Artificial Neural Networks.
[43] Maren, A., Harston, C. and Pap, R. (1991). Handbook of Neural Computing Applications.
[44] Himanen, V., Nijkamp, P., and Reggiani, A. (1998). Neural Networks in Transportation Applications.

- the symbol θ_j in f is the "bias", which determines the threshold value of the j_{th} neuron activation. Such a value is different for each neuron and can be considered as the weight of a "fictitious" connection, between an imaginary neuron and the j_{th} neuron. It provides a means of adding a constant value to the summed input, which can be useful to scale the overage input into a useful range (Maren)[45].

(5.1) $$A = f\left(\sum_{j=i}^{n} W_{ij} \cdot A_j + \theta_j \right)$$

5.2.2 Meso structure

The meso structure (Figure 5.2) is the architecture formed by all neurons; it is subdivided into layers and each layer contains a different number of neurons. Substantially, neurons are linked in three ways: a) "forward", when the direction of the neuron's connection is from the lower to the upper layer; b) "backward", when the direction is inverse with respect to the forward direction; and c) "lateral", when the neurons of a layer are connected (Maren)[45]:

The choice of an architecture and also of the way a network operates, depends on some considerations about the design of input and output representations. For example, some networks can work only with binary signals, others with real valued signals. Moreover, some can handle only static inputs, others can work with time varying sequences of information. With regard to the type of output, some networks produce only classification of input data, others may produce connections or adjustments in a process. All these distinctions can affect the designer's choice of a network, it is important to know the internal relations between input and output (Maren)[45].

---

[45] Maren, A., Harston, C. and Pap, R. (1991). Handbook of Neural Computing Applications.

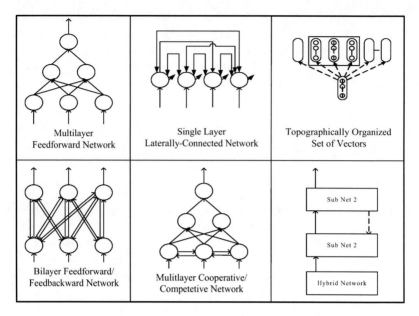

Figure 5.2   Six basic topologies of ANN meso structures (Himanen et al)[46]

5.2.3 Macro structure

The ANN macro structure is completely described by the connection of different meso structures. When no previously described single network can solve a problem, then it is useful to design a system of ANN, where each ANN "later" performs a specific task, interacting with others. Figure 5.8 shows an example of an ANN macro structure.

When an application has to be solved by means of ANN, it is important to choose an appropriate ANN design for its representation, in terms of its meso structure and considering the characteristics of individual neurons (micro structure). Once the architecture of ANN is defined, the subsequent phase concerns the learning process of ANN (Maren)[47].

5.3   Learning methodologies

The behaviour of an ANN depends mainly on the learning algorithm, since this phase provides the basic information on the environment at hand, by preparing the neural network to recognise and classify the patterns.

---

[46] Himanen, V., Nijkamp, P., and Reggiani, A. (1998). Neural Networks in Transportation Applications.
[47] Maren, A., Harston, C. and Pap, R. (1991). Handbook of Neural Computing Applications.

In particular during the learning phase, an ANN system may adapt or organise itself each time it examines a new sample, by changing the synaptic weights of each connection (the adjustment process). Consequently, since ANNs only learn a part of the possible samples (which can be infinite), the choice of the samples is very important (since the samples have to reflect the whole environmental pattern). Indeed, during the learning phase, internal representations of the samples take shape and define the patterns by means of vectors of real numbers, which can be modified in order to minimise the estimated error of the sample according to different criteria of numerical calculus (e.g., the mean squared error) (Himanen et al)[48].

In this context two different standard classes of methodologies of learning processes have been developed in the literature:
- Supervised;
- Unsupervised.

5.3.1 Supervised learning

This kind of learning implies that a network learns by examples, providing for every input configuration the respective output. The first network which made use of such an algorithm was the Rosenblatt perceptron (see Figure 5.3). It was made up of two layers: an input and an output layer (Himanen et al).

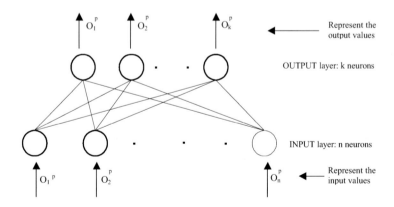

Figure 5.3   Perceptron architecture (Himanen et al)

Supervised learning incorporates an external teacher, so that each output unit is told what its desired response to input signals ought to be. During the learning process

---

[48] Himanen, V., Nijkamp, P., and Reggiani, A. (1998). Neural Networks in Transportation Applications.

global information may be required. Paradigms of supervised learning include error-correction learning, reinforcement learning and stochastic learning. An important issue concerning supervised learning is the problem of error convergence, for example the minimisation of errors between the desired and computed unit values. The aim is to determine a set of weights which minimises the error. One well known method, which is common to many learning paradigms, is the least mean square (LMS) convergence (Stergiou and Siganos)[49].

5.3.2 Unsupervised learning

Unsupervised algorithms distinguish themselves from the previous ones because they do not use a training set. As a consequence, they show both computational complexity and an inferior accuracy compared to the supervised algorithms; therefore, they are mainly useful in real time environments where a high execution speed is needed or whenever there is lack of information. Unsupervised training computes synaptic weights by examining how these modify their values according to the local information received from the neural signals (Himanen et al)[50].

Unsupervised learning uses no external teacher and is based upon only local information. It is also referred to as self-organisation, in the sense that it self-organises data presented to the network and detects their emergent collective properties. Paradigms of unsupervised learning are Hebbian learning and competitive learning (Stergiou and Siganos)[49].

5.4 Neural networks transfer functions

The behaviour of an ANN depends on both the weights and the input-output function (transfer function, f) that is specified for the units. This function typically falls into one of three categories (Stergiou and Siganos)[49]:

- ➤ Log-Sigmoid transfer function (see 5.4.1).
- ➤ Tan-Sigmoid transfer function (see 5.4.2).
- ➤ Linear transfer function (Purelin) (see 5.4.3).

For linear units, the output activity is proportional to the total weighted output. For threshold units, the output is set at one of two levels, depending on whether the total input is greater than or less than some threshold value. For sigmoid units, the output varies continuously but not linearly as the input changes. Sigmoid units bear a greater

---
[49] Stergiou and Siganos. (2001). Neural Networks.
[50] Himanen, V., Nijkamp, P., and Reggiani, A. (1998). Neural Networks in Transportation Applications.

resemblance to real neurones than do linear or threshold units, but all three must be considered rough approximations.

The following three transfer functions are the most commonly used transfer functions for backpropagation (MatLab)[51].

5.4.1 Log-Sigmoid transfer function

This transfer function is continuous, smoothly defined over the interval, monotonically increasing and differentiable (see Figure 5.4). This characteristic allows the formulation of backpropagation learning algorithms. Its algorithm is as shown in equation 5.2:

(5.2) $$f = \text{logsig}(n) = \frac{1}{1 + \exp^{(-n)}}$$

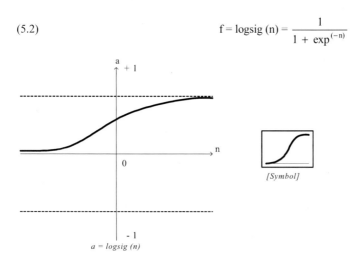

Figure 5.4   Log-Sigmoid transfer function (Matlab)

5.4.2 Tan-Sigmoid transfer function

This transfer function, is a squashing function of the form shown in Figure 5.5 that maps the input to the interval (-1,1). Its algorithm is as shown in equation 5.3:

(5.3) $$f = \text{tansig}(n) = \frac{2}{1 + \exp^{(-2 \cdot n)}} - 1$$

---

[51] Matlab (2001). The Neural Network Toolbox.

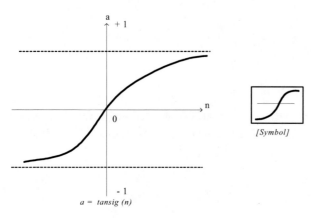

Figure 5.5   Tan-Sigmoid transfer function (Matlab)[52]

5.4.3   Linear transfer function (Purelin)

Is a transfer function (see Figure 5.6), that produces its input as its output. Its algorithm is as shown in equation 5.4:

$$(5.4) \qquad f = purelin(n) = n$$

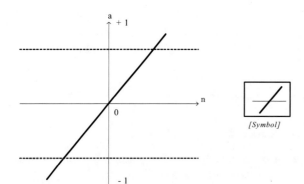

Figure 5.6   Linear transfer function (Purelin) (Matlab)

---

[52] Matlab (2001). The Neural Network Toolbox.

5.5 Work and functions of the artificial neural networks

Generally, neural networks are trained so that a particular input leads to a specific target output. Figure 5.7 illustrates the general functionality architecture of training. The network is adjusted, based on a comparison of the output and the target, until the output matches the target (Tsoukalas and Uhrig)[53].

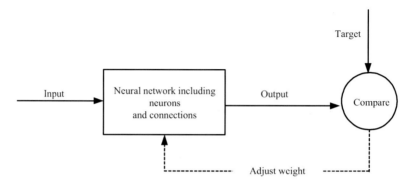

Figure 5.7   Neural network functionality

The neural network processing elements (neurons) are arranged in a sequence of layers with connections (weights) connecting the layers. Generally, there are three types of layers: input layer, hidden layer, and output layer. Typically, neural networks are fully connected, which means that all neurons in each layer are connected to each neuron in the next layer. Figure 5.8 shows the general architecture of a neural network (Haykin)[54]. In other words, the generic ANN structure is represented (see section 5.2).

---

[53] Tsoukalas, L.H., and Uhrig R. E. (1996). Fuzzy and Neural Approaches in Engineering.
[54] Haykin, S. (2nd ed.) (1999). Neural Networks: A Comprehensive Foundation

Artificial Neural Networks

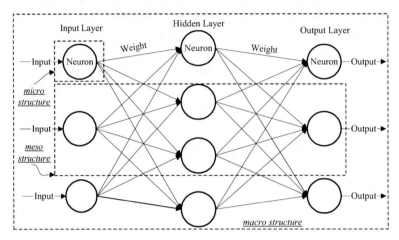

Figure 5.8   General neural network architecture

The processing elements (neurons) process the data before transferring it to the next layer and all the neurons function simultaneously in a parallel mode. The neurons receive input signals from the previous layer after being modified by weights. The neurons have two parts. The first part sums the weighted input and the second part applies an activation function on the input. Figure 5.9 shows the architecture of a typical neuron where X_i is the neuron input, W_{ij} is the weight from i_{th} layer to the j_{th} layer, and y_j is the neuron output (Tsoukalas and Uhrig)[55].

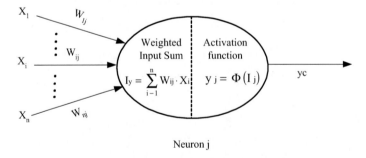

Figure 5.9   Neuron architecture (Tsoukalas and Uhrig)[55]

The activation function is the function that each node applies to the input to compute its output. One of the most widely used activation functions is the Log-Sigmoid Transfer

---

[55] Tsoukalas, L.H., and Uhrig R. E. (1996). Fuzzy and Neural Approaches in Engineering.

function (see 5.4.1). The choice depends on the problem to solve; linear or non linear, minimum and maximum values for inputs and outputs, learning algorithm, etc. A type of activation function (see Neural networks transfer functions, section 5.4) is chosen. Because under this research work, the main problem is considered linear (Risk-Factors and Total risk), the activation function purelin (see section 5.4.3) was selected to do the work (see chapter 8).

The difference between figure 5.1 and figure 5.9 is, that figure 5.1 shows only how a neuron (A_j) receives the outputs (A_n) and weights (W_{nj}) from other neurons and as well shows the outputs generated from neuron A_j (W_{jm} A_j). In the other side, figure 5.9 shows how the outputs (X_n W_{nj}) from other neurons coming to Neuron j are integrated in a sum (Weighted Input Sum) and then transformed into the limits of the selected activation function (y_j) to conform the output (yc) of Neuron j.

5.6 Backpropagation algorithm

A popular artificial neural network model is the back propagation network algorithm. The basic back propagation model is a three-layered forward architecture. Each layer contains a group of nodes that are linked together with nodes from other layers by connections among the nodes. Layers are connected only to the adjacent layers. The network is a feed-forward network, which means a neuron's output can only originate from a lower level, and a neuron's output can only be passed to a higher level. Figure 5.10 illustrates the basic structure of a feed forward neural network (Haykin)[56].

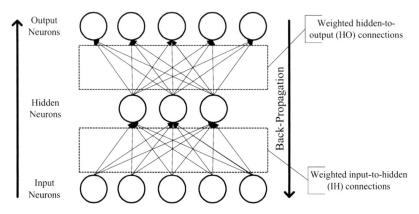

Figure 5.10   Back-propagation neural network

---

[56] Haykin, S. (2nd ed.) (1999). Neural Networks: A Comprehensive Foundation.

Backpropagation was created by generalizing the Widrow-Hoff learning rule to multiple-layer networks and non-linear differentiable transfer functions. Input vectors and the corresponding target vectors are used to train a network until it can approximate a function, associate input vectors with specific output vectors, or classify input vectors in an appropriate way as defined by the user. Standard backpropagation is a gradient descent algorithm, as is the Widrow-Hoff learning rule, in which the network weights are moved along the negative of the gradient of the performance function. The term back-propagation refers to the manner in which the gradient is computed for non-linear multiplayer networks. Properly trained backpropagation networks tend to give reasonable answers when presented with inputs that they have never seen. Typically, a new input leads to an output similar to the correct output for input vectors used in training that are similar to the new input being presented. This generalisation properly makes it possible to train a network on a representative set of input/target pairs and get good results without training the network on all possible input/output pairs (MatLab)[57].

As extra information regarding the Widrow-Hoff learning rule, the following description is mentioned:

Widrow-Hoff learning rule - A learning rule used to train single-layer linear networks. This rule is the predecessor of the backpropagation rule and is sometimes referred to as the delta rule (MatLab)[57]. The Widrow-Hoff learning rule operates with a single linear neuron model. Its summarized description is as follows (Haykin)[58]:

1. Initialization. Set

    $W_A(1) = 0 \quad \text{for } A = 1, 2, ..., p$

2. Filtering. For time $n = 1, 2, ...$, compute

    $$O(n) = \sum_{j=1}^{p} W_j(n) X_j(n)$$

    $e(n) = d(n) - O(n)$
    $W_A(n+1) = W_A(Lr) + ne(n) X_A(Lr) \quad \text{for } A = 1, 2, ..., p$

Figure 5.11 Summary of the Widrow-Hoff algorithm (Haykin)

---

[57] Matlab (2001). The Neural Network Toolbox.
[58] Haykin, S. (1994). Neural Networks, a comprehensive Foundation.

Where:
- W = weights
- A = name or number of weights
- P = set of values of the weight
- O = output signal
- N = iteration or discrete time
- J = neuron number
- X = individual signals produced by these sensors
- E = error signal
- D = desired target
- Lr = learning rate

For a detailed study of the Widrow-Hoff algorithm, the reader is referred to Haykin [59], and Widrow and Stearns [60].

There are many variations of the backpropagation algorithm. The simplest implementation of backpropagation learning updates the network weights and biases in the direction in which the performance function decreases most rapidly, the negative of the gradient.

One iteration of this algorithm can be written as shown in equation 5.5:

(5.5) $$X_k + 1 = X_k - \alpha_k \cdot g_k$$

Where X_k is a vector of current weights and biases, g_k is the current gradient, and α_k is the learning rate. There are two ways in which this gradient descent algorithm can be implemented: incremental mode and batch mode. In the incremental mode, the gradient is computed and the weights are updated after each input is applied to the network. In the batch mode all the inputs are applied to the network before the weights are updated (Matlab)[61].

For a practical view of how the Backpropagation Algorithm works, refer to the next section 5.7.

---

[59] Haykin, S. (2nd ed.) (1991). Adaptive Filter Theory.
[60] Widrow, B., and Stearns, S. D. (1985). Adaptive Signal Processing.
[61] Matlab (2001). The Neural Network Toolbox.

5.7 Example of the backpropagation algorithm

As a matter of clarifying how the backpropagation algorithm works, a simple example is carried out in this section. This network has two inputs, three hidden neurons and one output neuron. The network is shown in Figure 5.12.

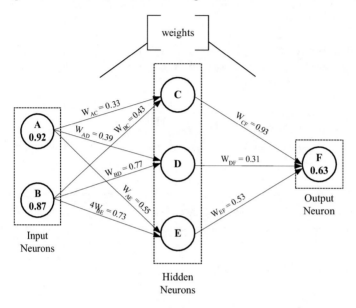

Figure 5.12   Sample network

5.7.1 Forward Propagation

A name and a value was proposed for each neuron, for example for neuron A the value equals to 0.92 and for neuron B the value is 0.87 (see Figure 5.12). A name is given as well for each of the hidden neurons (in this case 3 hidden neurons, see Figure 5.12), these names are C, D and E. Also the proposed weights (W) are settled for each connection between input and hidden layers and between hidden layers and output neuron. The output neuron is named F.

The values of the input neurons will be multiplied by their weights on the connecting links. So the input signal (S_{in}) for the hidden neuron C equals to:

$S_{in,AC,prim} = W_{AC} \cdot S_{out,AC,prim} = 0.33 \cdot 0.92 = 0.30$
$S_{in,BC,prim} = W_{BC} \cdot S_{out,BC,prim} = 0.43 \cdot 0.87 = \underline{0.37}$
$$SS_{in,C} = 0.67$$

The nomenclature "prim" means, that the input signal it has not yet transferred. In the other side, "S_{out}" denotes the output from the precedent neurons. While "S_{in}" is the input of the hidden neuron: the summation $SS_{in,C}$, is the total input of the hidden neuron (in this case C). Then, while using the formula of the Log-Sigmoid transfer function (see section 6.4.1), the output (change from input to output because the input values are transferred by the function) for the hidden neuron C is calculated.

(5.6)
$$A = \frac{1}{1 + \exp^{(-sum)}}$$

$$A = \frac{1}{1 + \exp^{(-0.67)}}$$

$$A = \frac{1}{1 + 0.51} = 0.66 \text{ output of the hidden neuron C}$$

Then, the value of 0.66 is the output for the hidden neuron C. The same procedure is carry out for the hidden neurons D and E. For the hidden neuron D we have:

$S_{in,AD,prim} = W_{AD} \cdot S_{out,AD,prim} = 0.39 \cdot 0.92 = 0.36$
$S_{in,BD,prim} = W_{BD} \cdot S_{out,BD,prim} = 0.77 \cdot 0.87 = \underline{0.67}$
$$SS_{in,D} = 1.03$$

Then, while using the formula of the Log-Sigmoid transfer function (see formula 5.6), the output of the hidden neuron D is calculated:

$$A = \frac{1}{1 + 0.36} = 0.73 \text{ output of the hidden neuron D}$$

Then, the value of 0.73 is the output for the hidden neuron D. The same procedure is carry out for the hidden neuron E. For the hidden neuron E we have:

$S_{in,AE,prim} = W_{AE} \cdot S_{out,AE,prim} = 0.55 \cdot 0.92 = 0.51$
$S_{in,BE,prim} = W_{AE} \cdot S_{out,BE,prim} = 0.73 \cdot 0.87 = \underline{0.63}$
$$SS_{in,E} = 1.14$$

76 Artificial Neural Networks

Then, while using the formula of the Log-Sigmoid transfer function (see formula 5.6), the output of the hidden neuron E is calculated.

$$A = \frac{1}{1+0.32} = 0.76 \text{ output of the hidden neuron E}$$

Then, the value of 0.76 is the output for the hidden neuron E.

In Figure 5.13, it is possible to see the output values for each hidden neuron. Now we are going to calculate the input value for the output neuron (only one in this example). Notice that every hidden neuron (see Figure 5.13) is connected to the output neuron. The value for the neuron in the output layer is calculated by the summation of each hidden layer neuron multiplied by their corresponding weight values (W).

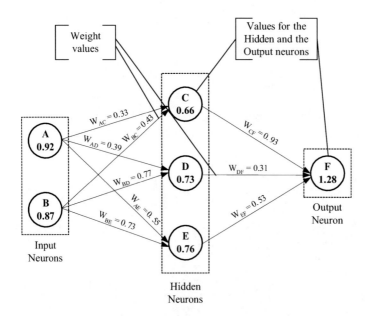

Figure 5.13   Outputs of the hidden neurons

So we have:
$S_{in,CF} = W_{CF} ? S_{out,CF,trans} = 0.93 \cdot 0.66 = 0.61$
$S_{in,DF} = W_{DF} ? S_{out,DF,trans} = 0.37 \cdot 0.73 = 0.27$
$S_{in,EF} = W_{EF} ? S_{out,EF,trans} = 0.53 \cdot 0.76 = \underline{0.40}$
$$S\ S_{in,F} = 1.28 \text{ actual value}$$

The term "trans" denotes that the values have been transferred. The actual value ($SS_{in,F}$) of the output neuron (see Figure 5.13), is compared to the "desired output". If different, an error value is calculated. For this example, the value of the "desired output" was proposed to be 0.63. So the error value is:

Actual value	= 1.28
Desired output	= 0.63
Error (Difference)	= 0.65

5.7.2 Backpropagation

The error value is back-propagated and used to modify the weights on each link connected to the output neuron. First of all, in order to make this adjustment of the weights. A learning rate needs to be chosen. For this case, the proposed value of the learning rate is 0.25. For this calculation, the hidden neuron values are multiply by the error value and by the learning rate:

The formula is as follows:

(5.7) $\qquad W_{AB} = N \cdot e \cdot L$

where:
W_{AB} = weight adjustment from neuron A to neuron B
N = neuron value
e = error for the output neuron
L = learning rate

Weight adjustment (W_{CF}) for neuron C:
$W_{CF} = 0.66 \cdot 0.65 \cdot 0.25 = 0.11$
$W_{new, CF} = W_{old} + W_{CF} = 0.93 + 0.11 = 1.04$

Weight adjustment (W_{DF}) for neuron D:
$W_{DF} = 0.73 \cdot 0.65 \cdot 0.25 = 0.12$
$W_{new, DF} = W_{old} + W_{DF} = 0.31 + 0.12 = 0.43$

Weight adjustment (W_{EF}) for neuron E:
$W_{EF} = 0.76 \cdot 0.65 \cdot 0.25 = 0.12$
$W_{new, EF} = W_{old} + W_{EF} = 0.53 + 0.12 = 0.65$

The purpose of the new weights is to update the connections between the neurons in order to reduce the error. In other words, a "New weight" is the actualised value of the

"Old weight", this is done with the weight adjustment (formula 5.7). The new weight values between the hidden neurons and the output neuron will be used in the next Forward Propagation (see Figure 5.14). Now, the error for each hidden layer is calculated. The error value is calculated by multiplying the error of the output neuron (0.65) by the respective weight that connects it to the hidden neuron we are working with. The formula is as follows:

(5.8) $$e_{hn} = e_{on} \cdot W_{hn}$$

e_{hn} = error in hidden neuron
e_{on} = error in output neuron
W_{hn} = weight in hidden neuron

For <u>hidden neuron C</u> the error is:
$e_C = 0.65 \cdot 0.93 = 0.60$

For <u>hidden neuron D</u> the error is:
$e_D = 0.65 \cdot 0.31 = 0.20$

For <u>hidden neuron E</u> the error is:
$e_E = 0.65 \cdot 0.53 = 0.34$

These error values are used to modify the weights from the hidden neurons to the input neurons. This modification will provide the new weight values between the input and the hidden neurons, which need to be used in the next Forward Propagation (see Figure 5.14).

Artificial Neural Networks 79

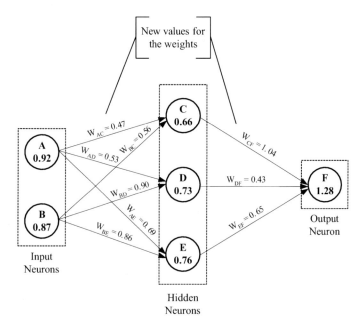

Figure 5.14   New weight values

For doing this modification, in other words for adjusting the weights from the hidden neurons to the input neurons, we multiply the value of each input neuron by the error value (from the hidden neurons) times the learning rate. In this case, the proposed learning rate is 0.13 (the learning rates used in different layers do not need to be the same). It is sometimes advantageous to use different learning rates. So we have:

The formula is as follows:

(5.9) $W_{AB} = I \cdot e \cdot L$

where:

W_{AB} = weight adjustment from neuron A to neuron B
I   = value of input neuron
e   = error from the hidden neuron
L   = learning rate

In <u>neuron A</u>, the weight adjustment (W_{AC}) is:
$W_{AC} = 0.92 \cdot 0.60 \cdot 0.13 = 0.07$

In <u>neuron A</u>, the weight adjustment (W_{AD}) is:
$W_{AD} = 0.92 \cdot 0.20 \cdot 0.13 = 0.02$

In <u>neuron A</u>, the weight adjustment (W_{AE}) is:
$W_{AE} = 0.92 \cdot 0.34 \cdot 0.13 = 0.04$

The SW_{Ai} of the weight adjustments for neuron A equals to $SW_{Ai} = (0.07 + 0.02 + 0.04) = 0.13$, so the new weights are:

$W_{new, AC} = W_{old} + SW_{Ai} = 0.33 + 0.14 = 0.47$
$W_{new, AD} = W_{old} + SW_{Ai} = 0.39 + 0.14 = 0.53$
$W_{new, AE} = W_{old} + SW_{Ai} = 0.55 + 0.14 = 0.69$

In <u>neuron B</u>, the weight adjustment (W_{BC}) is:
$W_{BC} = 0.87 \cdot 0.60 \cdot 0.13 = 0.07$

In <u>neuron B</u>, the weight adjustment (W_{BD}) is:
$W_{BD} = 0.87 \cdot 0.20 \cdot 0.13 = 0.02$

In <u>neuron B</u>, the weight adjustment (W_{BE}) is:
$W_{BE} = 0.87 \cdot 0.34 \cdot 0.13 = 0.04$

The SW_{Bi} of the weight adjustments for neuron B equals to $SW_{Bi} = (0.07 + 0.02 + 0.04) = 0.13$, the new weights are:

$W_{new, BC} = W_{old} + SW_{Bi} = 0.43 + 0.13 = 0.56$
$W_{new, BD} = W_{old} + SW_{Bi} = 0.77 + 0.13 = 0.90$
$W_{new, BE} = W_{old} + SW_{Bi} = 0.73 + 0.13 = 0.86$

With these new weight values (see Figure 5.14), the process is repeated in cycles (Forward and Back Propagation) until the "desired output" is reached by the training and the weight adjustments. Normally the error value between the "desired output" and the "actual output" will be zero or near.

5.8 The Matlab programme

At the present time there are several programmes (like Atree 3.0, Braincel, BrainMaker, Matlab, NeuroShell, NeuroSolutions, Trajan 3.0 Professional, etc) available in the software market that offer a software package for creating and applying Artificial Neural Networks. For this research work, the Matlab program was selected to work with Artificial Neural Networks because of its popularity and trustworthiness.

It is possible to work in two manners; one is with the command window and the second is with the Network/Data Manager window. The first one is more traditional orientated towards the use of Matlab while using its program commands. However, it can take more time to learn the necessary commands to create an ANN.

With the second option of using Matlab, the procedure for addressing the necessary steps to create an ANN is straightforward. This is, because all the possible options (training, input data, output data, simulation, etc.) are linked to a window where the data is managed and created. The use of this graphical tool saves time and creates a friendly context for the user.

The "Neural Network Tool Box", available with the Matlab program is perhaps not the most sophisticated option to create neural networks with a software program; but it is complete and reliable and allows the creation of different types of neural networks. In this research work, the use of the "Backpropagation Algorithm" was selected because it is one of the most popular ones and as well, has a potential record of success application within different research areas in the industry (for example in medicine, civil engineering, aerospace, etc).

While using the "Neural Network Tool Box", a general screen is used and is referred as the central point for the creation, design, testing and simulation, etc, of the neural network. The screen is called the Graphical User Interface (GUI) (see Figure 5.15). The GUI offers an easy way to work with neural networks and to check and control any progress on its performance. Also the GUI, is a bridge to the command lines of the program. Figure 5.15 shows the principal window of the GUI.

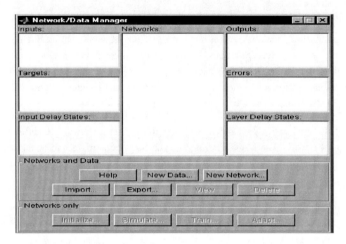

Figure 5.15   Graphical User Interface (GUI)

The data can be saved within the Matlab program and later be imported into the GUI. Also, the data can be stored on a diskette. The steps for saving, importing and exporting the data are quite simple and take no considerable amount of time.

The Matlab program provides a strong and effective mode using neural networks. While using the guidelines provided by the Matlab program, plus a previous study about neural networks, the user will find it relatively easy to work with the "Neural Network Tool Box".

6 Applications of Artificial Neural Networks in Civil Engineering

The examples presented under this section, represent the application of ANNs in Civil Engineering under different parameters (type of project, company task, etc.).

6.1 Modelling cost-flow forecasting for water pipeline projects using neural networks (A. H. Boussabaine, R. Thomas & T.M.S. Elhag)[62]

6.1.1 Project outline

This paper furthers work that already exists in the use of artificial intelligence techniques to forecast the cost-flow for construction projects. The paper explains the need for cost-flow forecasting and investigates the methods currently used to perform such a task. It introduces neural networks as an alternative approach to the existing methods. The relationship between the number of nodes used and the accuracy of the neural network in modelling an optimal solution is proposed for the case and a prototype system is developed (A. H. Boussabaine, R. Thomas & T.M.S. Elhag).

6.1.2 Conclusion

This paper reviewed the methods used for cost-flow forecasting and discussed their shortcomings. All of the existing models are based on linear regression where the best fit is sought. In these models, the factors that determine the shape of S-curves are very difficult to quantify and may not lend themselves to curve fitting since the needed representation cannot nicely fit into a quantitative description. Hence, it is impossible to perform such multi-attribute non-linear mappings by using regression models. For this type of environment neural networks are superior to the statistical models for forecasting project S-curves. Therefore, this work proposes an ANN system for solving this multi-attribute problem (A. H. Boussabaine, R. Thomas & T.M.S. Elhag).

Even so that ANN can reach acceptable (emulation of real targets) results, the main discussion will be located at the explanation of them. ANN uses at their middle structure (hidden layers) the so called "black box", which does not enable ANN to provide reasonable explanations about the results. Nevertheless, this disadvantage of the ANN can be overpassed when the ANN model is well structured (logical and sound mathematical relation between the input and output data, see section 7.10). The author of this work, did not include an analysis regarding the behaviour and the relationship between the input and ouput data vectors (see Figure 7.14), which is fundamental for providing more

---

[62] Boussabaine, A.H., Thomas, R., & Elhag, M.S. (1999). Modelling cost-flow forecasting for water pipeline projects using neural networks.

reliable results. Only the relationships between the number of nodes used and the accuracy of the neural network were closely examined.

The work does not show an analysis and impact of relevant variables which can affect seriously the cost flow of water pipeline projects. For example reasons as: lower project productivity (lower project performance), bad weather conditions, unstable economic conditions are just some of them. Other situations that as well affect the cost flow are: the project type, the contractual agreement, and the agreements for paying the contractor. Interesting will be that the author would be able to do a NPV analysis throughout the project life. This extra work, can provide a wide panorama of the cost-flow of this type of projects, and can enable as well the ANN to compare its results along and in each step of the project life cycle.

The results obtained in this work while using ANN do not provide better answers than the results obtained with traditional methods. The technique Cost-S-Curves has proved to offer still more reliable results. Perhaps, the reason of obtaining not acceptable results was that the data amount to train and test the ANN model was very limited. This experience shows that in order to further develop this type of research, more data will be require in order to possibly obtain better results.

6.2 Application of artificial neural network to forecast construction duration of buildings at the pre-design stage (S. Bhokha & S. O. Ogunlana)[63]

6.2.1 Project outline

The application of an artificial neural network (ANN) to forecasting the construction duration of buildings at the pre-design stage is described in this paper. A three layered back-propagation (BP) network consisting of eleven input nodes has been constructed. Ten binary input nodes represent basic information on building features (i.e. building function, structural system, foundation, height, exterior finishing, quality of interior decoration, and accessibility to the site), and one real-value input represents functional area. The input nodes are fully connected to one output node through hidden nodes.

6.2.2 Conclusion

An ANN designed and built to forecast the construction duration of building at the pre-design stage has been described in this article. The eleven independent variables of building features were selected based on their low covariance among one another. An attempt was made to establish a relationship between forecasting errors from the network and project duration. This shows that the network is equally good at forecasting the duration of both long and short duration projects (S. Bhokha & S. O. Ogunlana).

---

[63] Bhokha, S, & Ogunlana, S. O. (1999). Application of artificial neural network to forecast construction duration of buildings at the predesign stage.

The model presented shows the ANN structure selected for simulating time in construction projects. The variables (inputs) chosen seems to be realistic in the way of affecting the project performance. However, there are three inputs (functions, structures and exterior finishing) which are confusing because each of these inputs are represented by two input neurons at the input layer, instead of only one input neuron.

It is important to mention, that the work does not included any input-output relationship, which would be important for providing the logical sequence (input and output vectors, see Figure 7.14) of the ANN behaviour. By the inclusion of the input-output relationship, the model can be monitored while analysing how each input affects the final output; in addition, it is possible to decide whether the model has enough inputs variables or vice versa, it is necessary to reduce the number of them. In ANN, the increment of inputs, it is necessary then to increment the number of data to train the ANN.

The author do not clarified the difference used to describe the input data. The author uses "binary input nodes" and "one real value", without giving any definitions of each type of inputs. As well, there is a different scale used for evaluate the inputs values. These difference in the input structure, cause a confusion because implies that some inputs are more relevant to the ANN and if this is true, there is no explanation about it.

The results obtained with the model show a considerable difference between the simulated and the desired outputs. There is an average variation of 13.6 % between desired and simulated outputs. In other words, this difference can cause that in project in which this ANN is decided to be used, the duration obtained by the simulation might have with the real duration a difference up to 13.6 % in time. The reality is that while offering this difference, not many contractors will be whiling to use this ANN model because the forecasted duration of a project it may have a variance of 13.6 %, causing this a risk of failure to achieve the project on time

6.3 Neural network model for contractor's prequalification for local authority projects (F. Khosrowshahi)[64]

6.3.1 Project outline

The way in which clients or their consultants undertake the selection of contractors to tender for a given project is a highly complex process and can be very problematic. This is also true for public authorities as, for them, 'compulsory competitive tendering' is a relatively new concept. Despite its importance, contractor's prequalification is often based on heuristic techniques combining experience, judgement and intuition of the decision maker. This, primarily, stems from the fact that prequalification is not an exact science. For any project, the right choice of the contractor is one of the most important decisions that the client has to make. Therefore, it is envisaged that the development of an effective decision support model for contractor prequalification can yield significant benefits to the client. By implication, such a model can also be of considerable use to contractors: a model of this nature is an effective marketing tool or contractors to enhance their chances of success to obtain new work. To this end, this work offers a decision support model that predicts whether or not a contractor should be selected for tendering projects. The focus is on local authorities because, in the absence of a viable universal selection system, there are significant variations in the way they conduct prequalification (F. Khosrowshahi).

6.3.2 Conclusion

The qualification output (1) and the disqualification output (0) presented by the ANN are quite acceptable when compared with the actual outputs. The model has shown the potential of using an ANN as a tool or approach to be used for selecting the most suitable contractors at the tendering stage. Having in mind that the prequalification attributes considered for choosing the contractor are flexible and new ones can be added to the ANN model.

While distinguishing between public and private clients, this work has identified the benefits associated with client's ability to make a rational and informed decision on short listing contractors for invitation to bid. In light of the above, the research set to develop an alternative model for predicting the suitability of contractors to tender. The model is intended for use by public clients as well as proactive contractors who seek added knowledge about a client's decision process. This model can be used as an effective marketing tool by focusing on the factors that are perceived to be important to clients and by examining their chances of success to prequalify.

---

[64] Khosrowshahi, F. (1999). Neural network model for contractors' prequalification for local authority projects.

This model is a good endeavour to qualify contractors at the tender phase. Especially because of the wide range of attributes or factors involved in each decision that the client must keep in mind while deciding over the winner of the contract.

The ANN model in this work has shown acceptable results at the training and testing phases. There is no surety in assuming that the results will be satisfactory. There are examples of some ANN models used in construction projects where the desired target has not been reached. The reasons of this failure are several, however, the main reasons are: minimum training and testing data set, difficulty to emulate the real model, lack of explanation for the results provided by the ANN, lack of confidence on the results due to the black box of ANN and complication to design an ideal prototype to be used as an standard model.

In other words, there is no 100% security that an ANN will be the answer to our problem solving. The use of ANN implies a try and error procedure which perhaps cannot motivate some people interested more in probabilistic methods. It is a good technique; to design different structures of the ANN model and to train the model with different algorithms and to use unknown data for testing them. As alternative approach, it is good to compare the results obtained with the ANN with traditional methods or even do with results obtained with "rule of dumb". This will provide the ANN user with a sounded advice and information necessary to decide and make opinion over the usefulness and trustworthiness of the ANN model.

6.4 Artificial Neural Network for Measuring Organizational Effectiveness (Sinha and McKim)[65]

6.4.1 Project outline

An artificial neural network based methodology is applied for predicting the level of organisational effectiveness in a construction company. The methodology uses the competing value approach to identify 14 variables. These are conceptualised from four general categories of organisational characteristics relevant for examining effectiveness: structural context; person-oriented processes; strategic means and ends; and organisational flexibility, rules and regulations. Cross-sectional data has been collected from companies operating in institutional and commercial construction (Sinha and McKim).

6.4.2 Conclusion

A neural network for predicting organisational effectiveness of the construction company is developed. Fourteen organisational variables are identified and hypothesised to

---

[65] Sinha and McKim, (2000) Artificial Neural Network for Measuring Organizational Effectiveness

predict effectiveness. The network is a valuable practical tool that can give management of a construction company operating in institutional and commercial construction a pre-project view of their performance. To improve the performance of the network, significant variables through statistical analysis are identified, and then these variables are used in the input and hidden layers of the ANN. This has resulted in obtaining a simple and computationally efficient network that can predict organisational effectiveness of the construction firm (Sinha and McKim).

Although it is clear that the ANN approach tries to measure the organisational effectiveness of a construction company it is not clear how well the ANN approach behaves towards the actual outputs, which were used for training the network. The study only shows the errors for each model which are acceptable.

Further work in this area may look at comparing the predictive capabilities of the ANN based models with the predictive capabilities of the more traditional statistically based models.

6.5 River Stage Forecasting in Bangladesh: Neural Network Approach (Liong et al)[66]

6.5.1 Project outline

A relatively new approach, artificial neural network, was demonstrated in this study to be a highly suitable flow prediction tool yielding a very high degree of water-level prediction accuracy at Dhaka, Bangladesh, even for up to 7 lead days (Liong et al).

6.5.2 Conclusion

The application of an ANN has been successfully demonstrated on flow forecasting in Dhaka, Bangladesh. The flow prediction obtained has a very high degree of accuracy even for a 7-lead-day model. Although ANNs belong to a class of data-driven approaches, it is important to determine the dominant inputs, as this reduces the unnecessary data collection. Thus, the study demonstrates two results: (1) The suitability of an ANN for flow prediction with high accuracy at a fraction of the computational time required by the conventional rainfall-runoff models; and (2) a technique for detecting less sensitive input neurons in an effort to reduce unnecessary data collection and operational cost (Liong et al).

The ANN approach developed offers a good advantage over the traditional methods used for forecasting the stage on rivers. In this study, the ANN approach has probed to offer acceptable results and as well the adopted model (ANN structure) is logical and not complicated.

---

[66] Liong et al, (2000). River Stage Forecasting in Bangladesh: Neural Network Approach.

The problem with the conventional models is that they require a great deal of detailed data (e.g. topographical map, river networks and characteristics, soil characteristics, rainfall, and run-off data). Often, these data are not available. Conventional models are not ideal for real-time forecasting due to the large amount of detailed information they require and the associated long computational time.

6.6 Construction Labour Productivity Modelling with Neural Networks (Sonmez et al)[67]

6.6.1 Project outline

Construction labour productivity is affected by several factors. Modelling of construction labour productivity could be challenging when effects of multiple factors are considered simultaneously. In this paper a methodology based on the regression and neural network modelling is presented for quantitative evaluation of the impact of multiple factors on productivity. The methodology is applied to develop productivity models for concrete pouring, formwork, and concrete finishing tasks, using data compiled from eight building projects. The predictive behaviours of the models are compared with the previous productivities studies. Model results, advantages of the methodology, and study limitations are discussed (Sonmez et al).

6.6.2 Conclusion

Productivity models for concrete pouring, formwork, and concrete finishing tasks were developed using a methodology based on neural network modelling techniques. The use of neural networks helped the overall modelling process. The model results indicated that the effect of the factors on productivity may vary from task to task. Although some factors could have similar influences on productivity of a number of tasks, their rate of impact on productivity may be different. Improvements in modelling of construction productivity will hopefully lead to more realistic expectations and better planning decisions (Sonmez et al).

It is indeed a very good attempt to measure the impact of several factors into productivity, in this case for: concrete pouring, formwork and concrete finishing. Even that it is a small study, its results indicate that the effect of the factors on productivity may vary from task to task.

This approach might be usefully implemented in other construction processes (for example in bricklaying, excavation, foundations, etc.), and lead to more realistic and better planning decisions.

---

[67] Sonmez et al (1998). Construction Labour Productivity Modelling with Neural Networks.

The main impact of this study is that can provide feasible values of productivity rates which can be used for a better estimation of construction activities and costs.

6.7 General conclusions

The examples included of ANNs in civil engineering provide a brief panorama of the role and impact, as well the success or failure of this artificial intelligence tool.

It is risky indeed, at the beginning of a research project while using ANN to suppose a success. This is mainly because ANNs need a considerable amount of data. Not only for training the network, also for testing. In addition to this, it is extremely difficult to obtain real data from industrial partners.

The proposal to use ANNs in order to solve practical engineering problems is not a guarantee. The main purpose of ANN is to provide answers based on past similar situations; however, when the data used for training the network is not wide enough or is very limited, then the answers provide by the ANN perhaps will be not so accurate to the reality. Another critical problem that researchers can come up while using ANN; is that when the amount of data is small, then the ANN instead of simulating the output, the ANN will memorise the output. ANN required considerable time to learn about how they work, they are not straightforward and in order to get a sound knowledge, it requires a deep study and analysis over the theory of ANN first, and then it is possible to move on into the application field with the advise of people dedicated full time in working with ANNs.

ANNs represent an optional approach to solve engineering problems. In the field of civil engineering as has been demonstrate in this chapter, applications has been undertaken for implementing the possible solution of civil engineering problems while developing ANNs models. This section shows the application of ANNs in different areas of civil engineering, the present work will implement ANNs in the area of risk management. The success of any ANN model cannot fully mentioned until the model has been proved in enough cases in the industry. The present work will develop a theoretical part of the model and then, will implement the model while using real data. Nevertheless, the application will be done with very limited real data and for a specific contractor.

7 Theoretical Development of the Neuronal-Risk-Assessment System

7.1 Neuronal-Risk-Assessment System Definition

The Neuronal-Risk-Assessment System (NRAS) can be used as a systematic approach for quantifying, in terms of money, the risk involved in construction projects. The Neuronal-Risk-Assessment System can be considered a human-intuition approach which integrates the tools of *Artificial Neural Network* and *Risk Management* for the use and benefits of the contractor.

The main aim of the Neuronal-Risk-Assessment System is: *"to provide assistance to construction contractors in predicting the extra project cost (risk). This will assist the contractor in keeping capital expenditure and delivery time to predetermined values and takes necessary managerial action to avoid a shortage of cash, bankruptcy, and gives early warning of cost overruns"*.

In other words, the Neuronal-Risk-Assessment System will provide more realistic project costs because the risk value will be calculated based on past experience and on Risk-Factors (R-F). The approach used in this chapter is theory based; the application of it is provided in chapter 8. The Neuronal-Risk-Assessment System is a flexible tool that can be used within different types of construction projects and also within a variety of evaluation areas.

7.2 Objectives of the Neuronal-Risk-Assessment System

a) To offer to the contractor a practical and useful risk management tool, ready to be used at the bidding phase of a construction project. That means that the contractor will be able to use the NRAS for assessing the Total-Risk cost at the same time as other construction costs are calculated. As well, the contractor will have with the NRAS a check-list about risks, which can be updated at each tender.

b) To develop a practical neuronal model for predicting the Total-Risk cost in construction projects (refer to chapter 7 for the theoretical development of the model and to chapter 8 for its application and functional use). That means that the contractor will be able while using the NRAS to calculate the costs of risks for a specific project (at the tender phase) and to include this risk cost in the project's price given to the client.

c) Refine, write-up and disseminate the results of the study and develop further the innovative findings and strategies (refer to chapter 9). That means, to promote the use of the NRAS within the construction industry. As well to encourage further research on the topic while using the basic ideas of the NRAS with other type of Risk-Factors.

7.3 Working-plan of the Neuronal-Risk-Assessment System

The theoretical phases (from 1 to 4) of the NRAS are included in chapter seven; and the complete working-plan (all phases) will be applied in chapter eight. The phases are:

Phase 1 : Study of the existing situation: This phase involves the study and analysis of data of the projects chosen (total planned and total actual costs), in order to select the required data for further use together with the artificial neural network and the Risk-Factors (R-F). See sections 7.4 and 7.8.

Phase 2 : Project's Risk-Factors development: A list of the most important and harmful Risk-Factors (R-F) will be created. These Risk-Factors (see section 7.7) will be used as the input values for the artificial neural network and as the main contributors for the Total-Risk cost. A formula is proposed (see section 7.8.2) for assessing the Total-Risk value per project, representing the outputs values for the ANN.

Phase 3 : Risk-Factor and Total-Risk ranking: This phase involves the developing of a ranking scale for each Risk-Factor and Total-Risk values. A detailed explanation of the ranking scale will be carried out with due regard to the inputs and the outputs of the artificial neural network. See section 7.6.

Phase 4: Risk-Factor prototyping and structuring: The transformation of the input values and the output values (both obtained in phase 3) into the artificial neural network ranking-scale will be done with the help of the Risk Evaluation Form (section 7.8.3 and Figure 7.9). As well, the creation of an artificial neural network based on the Risk-Factors. See section 7.8.3 and Figure 7.10.

Phase 5: Artificial neural network training and experimenting phases: The creation, training and testing of different artificial neural networks (structure done in phase 4) is carried out. Also the analysis of the results is presented (to be done in chapter eight).

Phase 6: Validation of the system: The NRAS is tested. The testing process is carried out in different ways in order to evaluate the effectiveness of the system. All this is presented in chapter 8.

Phase 7: Refine and write up: This phase involves the writing up of the final conclusions, analysis, recommendations and the dissemination of the results (presented in chapter 9).

7.4 Actual Situation of the Problem: What is currently available?

Nowadays, a defined process or approach to determine the probable cost of the Total-Risk within a new project is not available. In the construction industry, one of the most challenging industries, the contractor faces the same difficult question for each project: How much I or we should put into the **"Total-Risk cost"** of this project?

In Germany, although a considerable amount of research work has been done within the risk management and the construction management areas, the necessity of an approach to satisfy the contractor's demand in calculating the value of the Total-Risk for a new project is still not satisfied. Nowadays, the value of the risk (Wagnis) and the profit (Gewinn) (W + G) assigned to a new project is commonly obtained by using formula 7.1. (Drees, Paul, 2002).

$$(7.1) \qquad W + G \ [€] = \frac{(W + G) \ [\%] \cdot AS \ [€]}{100}$$

Where:

$W + G \ [€]$ = Risk + Profit

$(W + G) \ [\%]$ = Risk and Profit value

$AS \ [€]$ = Offer Amount (Angebotssumme)

The assumption of using this formula to assess the risk and profit value is not logical and is not related to the most probable project risk factors. The main argument is that the formula proposed to use both concepts: Risk and Profit, does not provide a basis to justify the inclusion together of the Risk and Profit. In other words, the reason of the "Risk" is due negative events (risks), and in the other side, the reason of the "Profit" is due positive events (opportunities). The background of the "Risk" and the "profit" is different, the same happens with the factors that influence their values (Loss-Factors and Opportunity Factors). For that reason, this research work, focussed on proposing the use of the risk value as independent in the formula 7.1. In other words, it is proposed in this investigation to use formula 7.1 only when assessing the Profit (G) value, and when using the Neuronal-Risk-Assessment System for evaluating the Risk (W) value.

Actually in formula 7.1, the Risk and Profit values are based on a % of the project "Offer Amount "(see Table 7.1)"; where it is clear to see that there is no relation between the Risk and Profit, and the "Offer Amount". Therefore, this decision is not based on a system or on a process. The same value of the Risk and Profit (regularly used between 2 and 3 %) is usually given to every project, independent of its type or nature.

As mentioned previously, at the present time the values of the Risk and Profit are provided directly; in other words, without consideration of the project costs and characteristics. As support for the following discussion, the actual process for the calculation of the "Risk and Profit" (W+G) and the "Gross Offer Amount" (AS brutto) is shown in Table 7.1.

A definition of the basic elements used through the calculation process is described as:

- Single Costs of Each Piece of Work (Einzelkosten der Teilleistungen – EKT): These costs can be directly calculated for each part of work. The EKT calculate the amount of money for every individual piece of work.

- General Costs of the Construction Site (Baustellengemeinkosten – BGK): These general site costs can be directly added to a construction process. They are calculated separately and added in total to achieve the "Herstellkosten";

- Production Costs (Herstellkosten – HSK): Represent the costs related only to the construction site;

- General Business Costs (Allgemeine Geschäftskosten – AGK): The AGK costs are used for running a company, which cannot be added to a particular construction contract. However, they can be evaluated as a single operation or as a whole. They cannot therefore be directly added to a part of the bill of quantities;

- Total Production Costs (Selbstkosten – SK): Costs to the enterprise to build the project; including the general business costs;

- Risk and Profit (Wagnis und Gewinn – W + G): Estimate of the risk and the adaptation to the market by a planned profit;

- Net Offer Amount (Angebotssumme netto – AS netto): Market price; without value added taxes;

- Value Added Tax (Umsatzsteuer);

- Gross Offer Amount (Angebotssumme brutto – AS brutto): Price for the client.

Order	Phase	Type of Cost
	1	**Single Costs of the partial payment (EKT)**
	1.1	*Labour costs*
	1.1.1	Average wage
	1.1.2	Workers wage, Foreman wages
	1.1.3	Social costs
EKT	1.1.4	Incidental wage costs
	1.2	*Other costs*
	1.2.1	Material costs
	1.2.2	Preparation costs, scarf and trench-lining materials
	1.2.3	Business material costs
	1.3	*Devices costs*
	1.4	*Foreign payments costs*
	2	**General Costs of the Construction Site (BGK)**
	2.1	*Costs not dependable on time*
	2.1.1	Building-site set-up costs
	2.1.2	Building-site equipment
	2.1.3	Technical processing and control
	2.1.4	Construction uncertainty
+ BGK	2.1.5	Special costs
	2.2	*Costs dependable on time*
	2.2.1	Claim costs
	2.2.2	Business costs
	2.2.3	Construction management and workers costs
	2.2.4	General construction costs
= HSK	3	**Production Costs (HSK)**
+ AGK	4	**General Business Costs (AGK)**
= SK	5	**Total Production Costs (SK)**
+ (W + G)	6	**Uncertainty and Profit (W + G)**
= AS (netto)	7	**Net Offer Amount (AS netto)**
+ value added tax	8	**For example 16 %**
= AS (gross)	9	**Gross Offer Amount (AS brutto)**

Table 7.1   Calculation process of the "Gross Offer Amount" (AS brutto)

When the time comes to calculate the Risk and Profit (Table 7.1, Phase 6), there is no alternative method or approach to independently calculate the Risk (W). Instead of assessing the risk alone with its corresponding Risk-Factors as is proposed in this research work, a formula is given to calculate the Risk and Profit (W+G) together (Formula 7.1).

The purpose of Figure 7.1 is to show the confusing situation that occurs when formula 7.1 is used to calculate the Risk and Profit (W+G). Confusion occurs when the data concerning the AS (netto) is required beforehand rather than when it is really available. In other words, while observing Table 7.1 (phase 6) the formula requires the AS (netto). However, this (AS netto) is not available until phase 7. This can be easily calculated by using the percentage for risk and profit on the basis of total production costs. See Figure 7.1.

However it is not the central part of the research since the main aim is to separate the two important and different concepts: Risk (W) and Profit (G).

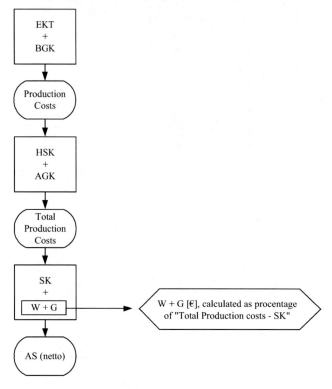

Figure 7.1   Calculating Risk and Profit by total production costs

Nevertheless, the method presented at Figure 7.1 for assessing Risk and Profit as basis of the total production costs does not solve the problematic situation mentioned in this research work, because it evaluates risk and profit together. For that reason, the work throughout this chapter is oriented towards an independent assessment of the values of risk (W) and profit (G).

7.5   Structure of the Neuronal-Risk-Assessment System

The structure of the system proposed for tackling the actual problem of assessing the Total-Risk is shown in Figure 7.2. This system integrates two different tasks carried out in a construction company. These tasks are: risk management and the calculation of construction prices. As well, it includes an artificial intelligence tool which has become very popular in the simulation of possible solutions for engineering problems. This tool is the artificial neural network.

The structure of the system as shown in Figure 7.2, indicates the connections between phases and is complemented by indicating the steps involved in each phase. For example for the risk analysis, risk evaluation and ANN design phase; steps 1 to 3 are involved. For the ANN training and testing phase, steps 4 to 5 take part. In the next phase, application phase, the corresponding steps are 6 to 8. Finally with the use phase the corresponding steps are 9 to 12.

In figure 7.3, information in detail is given for clarifying the activities involved in each step. For example, in step 3 named Neuronal Model Design. In order to do this, the creation of the ANN structure has to be done.

The three main components (risk management, artificial intelligence and the calculation of construction prices) of the Neuronal-Risk-Assessment System are composed of several activities. These activities are shown on Figure 7.3. A brief description of these components is included in the following sections.

7.5.1 Risk management

Under this section of the system, the Risk-Factors (see Table 7.2) created after several discussions with a Dresdner contractor are used in order to decide their classification and scores (see Table 7.3). Even though only 17 Risk-Factors were used to evaluate the project Total-Risk, there are other factors, called "Specific Risk-Factors" (see section 7.7.3) which are developed from the main 17 Risk-Factors.

It is important to clarify, that, for this work, only 17 Risk-Factors are used because these are the ones which represent the risk-profile of the contractor, who in this case has provided the data of 16 projects. Obviously, for another contractor, the number of Risk-Factors can be the same, smaller or even bigger.

The 17 Risk-Factors will represent the ANN inputs (a detailed explanation is given in section 7.7). The decision over the Risk-Factors was made by an experienced project manager, who has worked closely to the 16 projects used in this work.

Regarding the Total-Risk in this research work, the impact of all the Risk-Factors (17 in this case) represents the degree of the risk for a project. In order to determine the value of the Total-Risk, a formula (see section 7.8.2) was proposed as an alternative to obtain this value for each project. The formula will require the value of the actual profit and the actual value of the total production costs from each project.

98 Theoretical Development of the Neuronal-Risk-Assessment System

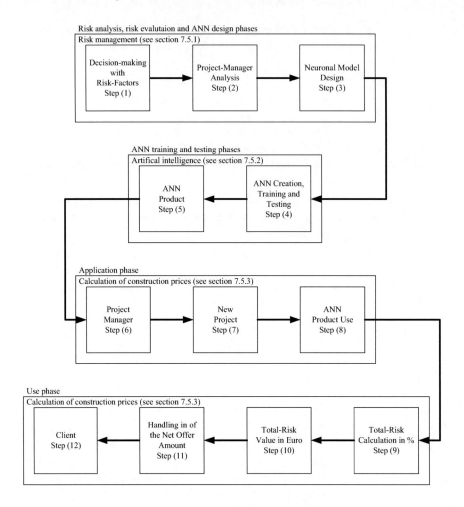

Figure 7.2   Structure of the Neuronal-Risk-Assessment System

It can be argued, that the meaning of this formula does not satisfy everyone, but because there is no formula at the moment, which can be used to calculate the Total-Risk value of a construction project, the consideration of this formula as a reasonable first attempt to formalise the calculation of project risk is encourage (a detailed explanation over this formula and its application is given in section 7.8.2).

The value of the Total-Risk obtained with each project, will represent the output to be used as the desired target at the output layer of the ANN. In this case, 16 values of the Total-Risk for 16 different projects.

Until here, the steps of decision-making with the Risk-Factors and the analysis made by the project manager have been covered. With these data, the inputs (Risk-Factors) and outputs (Total-Risk values per project), it is possible to advance to the next step which corresponds to the design of the structure of the ANN.

Firstly, a detailed explanation of the input and output ranking scales is given in section 7.6. The purpose of these ranking scales is to measure the impact of each Risk-Factor and the Total-Risk of each project. Afterwards, this value is transformed into the artificial neural network ranking value. See Figure 7.3 (steps 1, 2 and 3) for the detailed activities involved. This is done with the help of the Risk Evaluation Form (see Figure 7.9).

The neuronal model design, means deciding the structure of the ANN. However, there are two constants. The first one are the Risk-Factors which form the 17 inputs, and the second one are the 16 Total-Risk values which represent the output for each project.

The variable in this case, is the hidden layer of the ANN. For that reason, several different models are created. As a guide to a prototype of the ANN structure, refer to figure 7.10. Also, in order to see the 100 different neuronal models, see figure 8.5 in chapter 8.

For example, choosing a hypothetical project, the 17 Risk-Factors are evaluated as is done in Table 7.3. In chapter 8 (table 8.1), the same procedure is adopted but in this case for all the 16 projects. Regarding the Total-Risk value, the assessment is carried out as shown in section 7.8.2 (formula 7.2). In the example shown in section 7.8.2, the Total-Risk value corresponds to 4.98 % (0.0498 in ANN values). The same procedure is followed for all the remaining 16 projects, in this case the values of the Total-Risk per project are shown in table 8.3 (column 7) in chapter 8.

7.5.2 Artificial intelligence

This section is fully orientated to the creation, training and testing of the ANN. Complemented with obtaining a final product.

The structure of the ANN is design and based on the data-analysis (see section 7.8.2) for deciding the value of the outputs (Total-Risk), also it is based on the Risk-Factors (see section 7.7, and also table 7.2). With this data, the structure of the ANN can be built-up like is shown in Figure 7.10 and is complemented with section 7.8.3.

The creation of the ANN, as mentioned in the risk management section, is based mainly on the 17 inputs (Risk-Factors per project) and the 16 outputs (Total-Risk values per project). The extra step now is to decide a different order for training and testing the system, and also, deciding a different number of hidden neurons at the hidden layer. The whole proposal for creating, training and testing the NRAS can be seen in Figure 8.8 from chapter 8.

For example, Figure 8.5 in chapter 8 describes the testing procedure of the NRAS. By choosing for example the first one, in this case NRAS (A) model. The framework for testing means:

a) That the order of the 12 projects for training the system is: 1, 2, 3, 4, 5, 6, 7, 8, 9, 10, 11 and 12.

b) For the NRAS(A) model, 10 different ANNs will be created.

c) The difference between the networks created in the last point b), is the number of neurons selected for the hidden layer. For example: BP1, means a network trained with the projects order like in point a), but only one neuron in the hidden layer. Another example, in this case BP7, the order chosen for the projects is the same but now the network has 7 neurons in the hidden layer.

d) Steps a) to c) are done for each of the NRAS models shown in Figure 8.5. Achieving the creation of 100 different ANNs.

After each of the 100 ANN are trained. The next step is to test the reliability of the system. In this case by using the remaining four projects (13, 14, 15 and 16), which were not included in the training set. In the training set, only projects 1 to 12 were used. The results obtained with the test (simulation) are compared with the desired outputs of projects 13 to 16, and also the ANN reliability is evaluated while using the mean-square error (MSE). All this section is noted as the testing phase of the system (see Figure 8.5 in chapter 8).

The training and the simulation of each of the ten models, was made by using commercial software for ANNs. In this case, the Matlab programme was chosen. Refer to section 5.8 of chapter 5 for a detailed explanation of this software.

After the simulation for all the models was finished. The model which offers the best results is chosen to continue further with the analysis. In this case, the NRAS (A) model was the one which behaves better (see sections 8.5.2 and 8.5.3 for its results), and refer to appendix II for the results of the other models. In this case, the NRAS(A) model represents the final product. See Figure 7.3 (steps 4 and 5) for the detailed activities involved.

7.5.3 Calculation of construction prices

The activities under this section include the application of the final product (NRAS(A)), while using the results obtained at the simulation phase, against the actual values obtained without using the NRAS.

Each remaining project (13, 14, 15 and 16) was used to simulate their outputs with the NRAS (A), using also the "Risk Evaluation Form" (see Figure 7.9). The Risk Evaluation Form includes the Risk-Factors to be evaluated, obtaining with this the simulated value of the Total-Risk for each project, in this case for projects 13, 14, 15 and 16. See Figure 7.3 (step 6 to step 12) for the detailed activities involved.

The further analysis made with the NRAS (A) results are related to compare in money terms its meaning against actual values proposed with the actual approach to evaluated the Total-Risk (see section 7.4). The analysis is shown in section 7.11 (chapter 7) as the expected results and in sections 8.5.4 and 8.5.5 as the meaning and impact of the NRAS results and the risk management evaluation.

The activities from section 7.5.1 "Risk Management" and 7.5.2 "Artificial Intelligence", from Figure 7.3, represent the most important elements of the Neuronal-Risk-Assessment System.

From the three main sections (7.5.1, 7.5.2 and 7.5.3) described in Figure 7.3. In this chapter, the theoretical development of sections 7.5.1 and 7.5.2 will be carried out. For the risk management section (section 7.5.1) refer to sections 7.6, 7.7 and 7.8 from chapter 7. In chapter 8, refer to sections 8.2 and 8.3.

The sections which represent the artificial intelligence section (7.5.2) are represented in chapter 7 by sections 7.6, 7.7, 7.8, 7.9 and 7.10. In chapter 8, the sections which represent the artificial intelligence section are 8.5.1, 8.5.2, 8.5.3 and 8.4.

Finally, the section which represents the calculation of the construction prices (section 7.5.3) in chapter 7 is section 7.11. In chapter 8, the sections which represent the calculation of the construction prices are 8.5.4 and 8.5.5.

102　Theoretical Development of the Neuronal-Risk-Assessment System

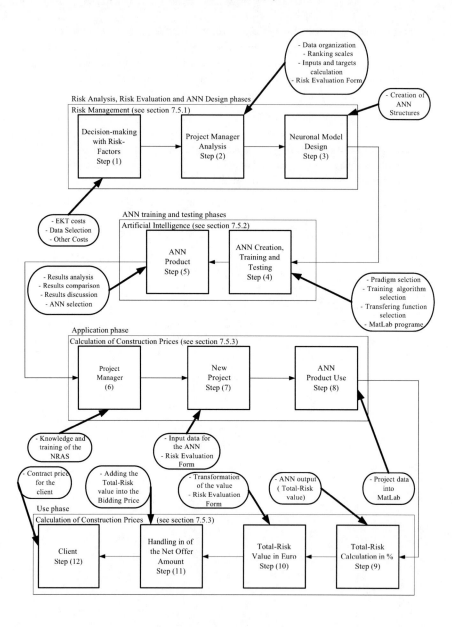

Figure 7.3　Basic activities of the Neuronal-Risk-Assessment System

7.6 Ranking scales of the Neuronal-Risk-Assessment System

Before the data is analysed the corresponding risk scores for each Risk-Factor are assessed. The way in which the impact of the Risk-Factors is classified, for the purpose of this research work is shown as a Risk-Factors (inputs) ranking scale (Figure 7.4).

This is very important as it will offer a clear idea of how the Total-Risk value together with its corresponding Risk-Factor impacts will be used. In other words, these ranking scales will help to identify and classify how "dangerous" each Risk Factor is for the project's profit.

Firstly, there is a necessity to limit the maximum and minimum values of the Risk-Factors. For this purpose, the ranking scale is built up with three main concepts: Risk-Factor Classification, Risk-Factor values in % (inputs scale) and Risk-Factors in Neuronal Scores. The meaning of these main concepts is as follow:

a) <u>Risk-Factor Classification:</u> Seven different classifications for the Risk-Factors are proposed, starting from the lowest one which corresponds to "Extreme Low" until the biggest one which corresponds to "Extreme high". This is just a proposal of the classification, however, for other risk managers perhaps this classification needs to be increased or decreased.

b) <u>Risk-Factor values:</u> Corresponding to each of the 7 classifications, a value is assigned in order to measure the Risk-Factor value in %. It is also called "Input Scale", because this scale will be used to assign a value to each of the 17 Risk-Factors used as can be seen in the Risk Evaluation Form (Figure 7.9). The scale chosen was from 0 to 100 %. Giving for each classification a accurate value, instead of having for example a scale of values (for example of 0 to 15 %) which will complicate the evaluation procedure for each Risk-Factor.

c) <u>Risk-Factors in neuronal scores:</u> The values used for scoring the Risk-Factors are related to the minimum and the maximum values allowed by the "purelin transfer function" (see chapter 6, section 6.4.3). With this neuronal scale, the values assigned to each Risk-Factor, will be transformed into neuronal values in a very easy way. For example, if a Risk-Factor has a "high classification", it corresponds a Risk-Factor value of 65 % and in neuronal values corresponds a value of .65.

This process will make it easier for using the Risk Evaluation Form with the project manager and then for feeding the simulation programme (Matlab in this case).

A similar procedure was adopted for creating the Total-Risk (outputs) ranking scale. The scale is form of three main concepts; these are: Actual Total-Risk classification, Actual Total-Risk value in % (output scale) and Total-Risk in Neuronal values.

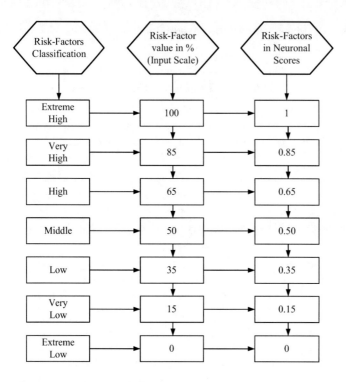

Figure 7.4   Risk-Factors (inputs) ranking scale

a) <u>Actual Total-Risk classification:</u> Seven different classifications are included for the Total-Risk values. As with the Risk-Factors, the lowest classification is "Extreme Low" and the biggest is "Extreme high".

b) <u>Actual Total-Risk value in %:</u> In this case, two values of the Total-Risk for each classification are provided rather that only one as with the Risk-Factors. In other words, a range of values is given in order to properly select the Total-Risk worth. For example, if the Total-Risk of a project is classified as very low, that means that the value in % of the Total-Risk equals from 11 to 20 % (minimum and maximum). While using these ranges, the possible percentage of the Total-Risk in a project is covered from 0 to 100 %.

c) <u>Total-Risk in neuronal values:</u> As well like in the Risk-Factors scale, the values used for scoring the Total-risk are related to the minimum and maximum values allowed by the "purelin transfer function" (see chapter 6, section 6.4.3). In a very easy way, a percentage value of the Total-Risk can be transformed into neuronal values. For example, if a project has a very low classification of the Total-Risk, and this Total-Risk equals to 11 %, that means that in neuronal values equals to .11.

The evidence of why the scales from Figures 7.4 and 7.5 have these particular classifications and values, is basically supported in the author knowledge about risk management. However, a basis guideline was founded in the research provided in CIRIA SP154 (2002), CIRIA 125 (1996) and in AS/NZS 4360 (1999).

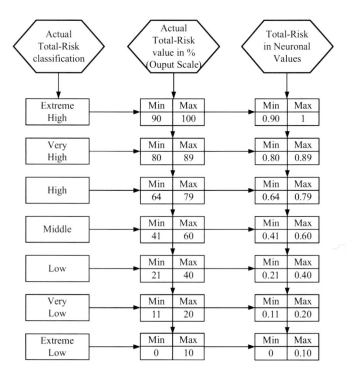

Figure 7.5   Total-Risk (outputs) ranking scale

7.7   Risk-Factors creation

7.7.1   Introduction

The main purpose of the development of Risk-Factors (Figure 7.3, step first) is to provide to the project evaluator an easy and logical way for assessing all the most important and relevant Risk-Factors, which can be implied at the starting phase of a construction project of all sorts.

The origin of the Risk-Factors came from the different areas of risk (political, economic, management, force majeure, etc. see Figure 7.6) enclosed in every construction process, where the risk impacts are commonly present. In other words, the Risk-Factors offer an opportunity to the project evaluator to assess social risks, operational risks, political risks, environmental risks, etc; all these in terms of money.

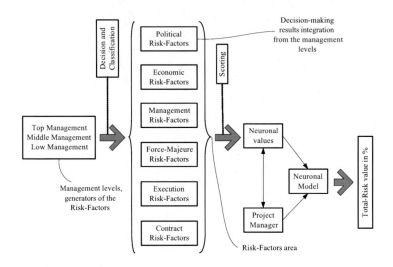

Figure 7.6   Purpose of the Risk-Factors

Using the Risk-Factors proposed in this work (see Table 7.2), the contractor will have the opportunity to evaluate the impact of the most common risks presented in a middle size contractor, from a developed country, and at the operational management level.

The reason of why only 17 Risk-Factors were created, is because after several discussions held with the contractor, the final decision was to concentrate on these 17 Risk-Factors. Obviously, for another types of contractor and perhaps in a developing country, the numbers of Risk-Factors and as well the type of Risk-Factors may be different.

It is extremely difficult to create a general guideline for the Risk-Factors in construction projects. The reasons are many, for example, the uniqueness of the construction projects, the geographical area, the management experience of the team, etc. Nevertheless, the Risk-Factors developed in the present work are satisfactory for this study.

7.7.2 Risk-Factors development

As mentioned in section 7.7.1, the Risk-Factor background represents the risk impact from different risk areas of the project's uncertainty. These Risk-Factors will be included in the "Risk Evaluation Form" which will be used as a practical tool for the project evaluator (see section 7.8.3 for a detailed explanation). Different areas (Environmental, social, economical, political, commercial, financial) and levels like (Top management, middle management and the operational level) will work together in this assessing process. See Figure 7.6:

Table 7.2 describes the proposed Risk-Factors. It might be argued that this Risk-Factors assessment framework might be limited or not wide enough to cover specific construction projects, for example oil platform, dams, etc. Nevertheless, this framework is an attempt to provide an extra help for the contractor in evaluating risks in construction projects.

Once the project evaluator has used this Risk-Factors assessment framework, the next step is to transform the results (Risk-Factors Score, see table 7.2 column 5) into neuronal values. This is done with the help of the "Risk Evaluation Form" (see section 7.8.3). It is important to mention that, with the help of these Risk-Factors it is possible to make a proper risk analysis (risk identification, analysis and control). At this phase (project start), it will be possible also to evaluate multiple risk scenarios that the project might deal with (See Risk Profile in chapter 8).

7.7.3 Specific Risk-Factors

In order to provide the reader a helpful explanation about what is the range of each Risk-Factor (Table 7.2), the following specific Risk-Factors were created.

The main reason of the development of the specific Risk-Factors is because it will be too complicated to include all of them in the Risk Evaluation Form (Figure 7.9). As well, after the discussion held with the contractor, there was an agreement to provide an explanation (using the Specific Risk Factors) of the scope of the Risk-Factors.

For example, with the first Risk-Factor called "Risk caused by the change in law, regulations, etc.", it is too complicated and not practical to describe all its range. For that reason, the reader can see its Specific Risk Factors corresponding to the first Risk-Factor, which in this case are five specific Risk-Factors. The same assumption is adopted for the rest of the Risk-Factors.

Risk-Factor Code	Risk-Factors	Risk-Factor Area	Risk-Factor Classification	Risk-Factor Score
R-F1	Risks caused by the change in law, regulations, etc.	Political		
R-F2	Risks by problems on permissions	Political		
R-F3	Inflation risks	Economic		
R-F4	Management risks caused by the subcontractor	Management		
R-F5	Contract risks caused by the subcontractor	Management		
R-F6	Risks from security and health protection	Management		
R-F7	Force majeure risks	Force Majeure		
R-F8	Weather risks	Force Majeure		
R-F9	Transport risks	Execution		
R-F10	Design and construction risks	Execution		
R-F11	Quality risks	Execution		
R-F12	Technical and execution risks	Execution		
R-F13	Risks from water and air pollution	Execution		
R-F14	Contract risks	Contract		
R-F15	Guarantee risks	Contract		
R-F16	Business and Market risks	Contract		
R-F17	Risks caused by the Client	Contract		

Table 7.2   Risk-Factors assessment framework

1) Risks caused by the change in law, regulations, etc

- ➢ Tariff changes during the construction period,
- ➢ Sound bureaucracy (late approvals),
- ➢ Cost increase due to changes of policies,
- ➢ Flexibility in the legal-political system (electoral time),
- ➢ All risks without such from traffic (see No. 9), environmental protection (see Nr. 13) and permission (see No. 2).

2) Risks by problems on permissions

- ➢ Complicated procedures for obtaining the permissions,
- ➢ Late approvals of permissions due to severe bureaucracy and the conditions likely to be imposed,
- ➢ Delays generated by corruption and bribery.

3) Inflation risks

- Change of currency or currency fluctuation,
- General inflation,
- Project cash flow affecting the rate of return (profit) of the project.

4) Management risks caused by the subcontractor

- Non-experienced subcontractors (shortage of construction engineering and management skills in a specific project),
- Insufficient financial resources (caused by external or internal economic catastrophes),
- Lack of effective leadership of the subcontractor.

5) Contract risks caused by the subcontractor

- Interface risks,
- Too short guarantee.

6) Risks from security and health protection

- Accidents caused by inappropriate system/ procedure for management of safety and security on site,
- Accidents caused by unskilled workers while using high-technology equipment,
- Accidents caused by a poor supervision/coordination (communication) between the construction manager and its subordinates (inappropriate system/ procedure for communication and office management),
- Accidents caused by external facts (for example social crisis in unstable countries),
- New construction methods,
- Accidents, caused by external facts (e.g. social crises in the unstable countries),
- Accidents, caused by bad supervision and co-ordination.

7) Force majeure risks

- Damages caused by unusual conditions such as: weather, soil/physical conditions in site, earthquakes, floods, fires, storms, embargoes, import and export restrictions, groundwater, etc.,
- Damages caused by environmental impacts (weather conditions disruption).

8) Weather risks
- Bad weather in sensitive building procedures (e.g. earthwork in rainy periods),
- Project delay caused by non creditable weather forecast,
- High possibility of adverse weather conditions during the construction period.

9) Transport risks
- Disruptions caused by inadequate traffic volume (inadequate traffic forecasts),
- "Right-of-way" disputes (e.g., archaeological mines),
- Geological risks (nature of site materials, groundwater),
- Restriction of space for using secondary roads,
- Incompetence of transportation facilities.

10) Design and construction risks
- Shortage of design skills (in-house, nationally, or internationally),
- Shortage of construction engineering and management skills,
- Lack of utilities and services (water, gas, electricity),
- Shortage of resources (human, material and equipment),
- Inappropriate system/ procedure for planning and schedule control,
- Inappropriate system/procedure for estimating and cost control.

11) Quality risks
- Poor quality of procured accessory facilities,
- Poor quality of procured materials,
- Stringent product quality requirements,
- Inconsistent grade quality throughout the ore body or deposit,
- Inappropriate system/ procedure for quality assurance/ quality control.

12) Technical and execution risks
- Untried process methods,
- Extreme operating conditions,
- Lack of equipment for hire, spare parts and repair facilities,
- Lack of training of the operating personnel,
- Low productivity/ unreliable and obsolete equipment,
- Design changes,
- Equipment failure,
- Materials shortage,
- No availability of operators and gang of workers,
- Low performance of the project output (productivity),
- Shortage supply of materials and other resources,

- Accidents on site,
- Obsoleteness of building equipment (high maintenance costs),
- Shortage in skilful workers,
- Use of different construction methods or processes,
- Incompetence of transportation facilities,
- Lack of training (for operator) in new technological equipment,
- Shortage of construction engineering and management skills,
- Extreme operating conditions,
- Lack of utilities and services,
- Lack of suitable construction materials and consumables,
- Lack of equipment for hire, spare parts and repair facilities,
- Lack of effective leadership,
- Excessive organisational hierarchy with strained lines of communication,
- Inflexible organisational structure incapable of adaptation to the required changes,
- Complex relocation of existing utilities/ facilities,
- Increase in site overheads,
- Accidents on site,
- Problems due to partner practise differences (different mode of using tools and equipment),
- High costs on equipment (rent of special equipment),
- Bad/ low productivity of workers,
- Demands of production space, construction access and handling of standard equipment,
- Demands of innovative construction methods,
- Training of the operating personnel and managers,
- Special provision for storage.

13) Risks from water and air pollution
- Improper use of natural resources,
- Use of sophisticated equipment that affects the natural environment,
- Use of perilous materials,
- Improper feasibility studies to determine/ formulate appropriate responses.

14) Contract risks
- Autocratic contract conditions,
- Adversarial elements (lack of an appropriate mediation/ arbitration provision),
- Divergence of contractual terms from project objectives,
- Complicated contractual terms, matters, procedures, etc., occupying inordinate amount of management time,

- Complicated joint venture structure,
- High risks regarding the used of materials (short contract time and periods etc., e.g. also with supplement permission).

15) Guarantee risks
- Long guarantee periods,
- Difficult acceptance procedures.

16) Business and Market risks
- Lack of effective leadership,
- Complicated decision-making process,
- Inflexible organisational structure incapable of adaptation to the required changes,
- Low demand of the market (inadequate forecast about market demand),
- Adverse publicity,
- Competition from other companies,
- Local protectionism,
- Bankruptcy of project partner,
- Fall short of expected income from project use,
- Payment delays caused by the Client.

17) Risks caused by the Client
- Planning changes,
- Delays in the project (caused by the Client),
- Cancellation or removal of the project,
- Difficult Client.

7.8   Data analysis

In this section, the theoretical procedure for obtaining the structure of the ANN, the value of the inputs and outputs will be carried out.

Beginning with section 7.8.1, a graphical comparison (Figure 7.7) is made between the actual approach used in Germany for obtaining the bidding price of a project and the ANN approach developed and proposed in this work. Figure 7.7 illustrates in general terms, each step followed by each approach.

Section 7.8.2 deals with a very important point, the calculation of the Total-Risk value per project; it is under this section, where formula 7.2 is proposed as an attempt to formalise and standardised the Total-Risk calculation in this work. A detailed explanation of the components of the formula, the analysis and the assessment of the Risk-Factors and Total-Risk is included.

Finally in section 7.8.3, a form is created in order to facilitate the evaluation of the Risk-Factors, this form is called "Risk Evaluation Form" (see Figure 7.9). As well, with the results obtained in section 7.8.2, the structure of the ANN is built up (see Figure 7.10).

The information used for making the calculations is fully provided by a "Dresdner construction company". The intention of this data analysis exercise is to show in detail each step involved in the analysis, calculation and transformation of the data until is ready to be used with the ANN. However, this example only deals with the cost-data from one project example. The same procedure is possible to be adopted with other types of cost-data. This theoretical procedure for the data- analysis is applied with a set of real cost-data in chapter eight; as a proof of the reliability and usefulness of the Neuronal-Risk-Assessment System and using the information from 16 projects.

7.8.1 Total planned costs, total actual costs and Risk-Factors integration

Figure 7.7 shows a more detailed process used in the Neuronal-Risk-Assessment System for the calculation of the project's actual Total-Risk.

The ANN approach is more realistic because with the actual approach, where the Profit and Risk are calculated together while using formula 7.1, the present work proposes to evaluate the Total-Risk, independently from the Profit. Refer to "W + G calculation" in the left side of Figure 7.7 to see the current approach and refer to "Total-Risk value (with NRAS)" and "+ Profit (G with formula 7.1)" in order to see how the ANN approach proposed a different path to calculate the Risk and the Profit independently.

This new process follows a more realistic and logical order because the risk is obtained independently from the "AS netto". In other words, there is a relationship between the Total-Risk value and the probable risks that cause this Total-Risk effect; it is completely different than the way generally carried out in Germany, where a number is given in % to assess the risk value, but nothing is given instead to justify this number (see Figure 7.7 (right side) to see the Risk-Factors and Total-Risk value relation). Instead of assigning an invariable percentage value without the nature, type and magnitude of the project; this approach provides more realistic values for the risk. In most cases the value depends on the type of project.

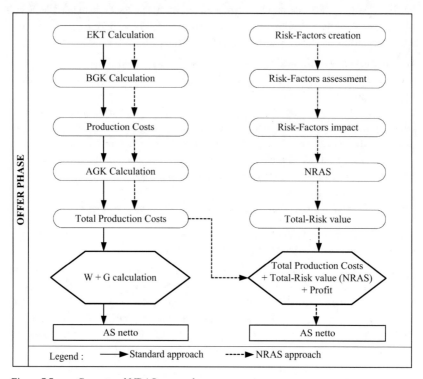

Figure 7.7   Current and NRAS approaches

7.8.2   Risk-Factors and Total-Risk analysis and assessment value

This section deals with a discussion focused on the analysis and assessment over the data used for the NRAS. The two main data-elements of the system are the Risk-factors and the Total-Risk value. It is important for the comprehension of the system, to illustrate the relationship existing between the Risk-Factors and the Total-Risk value. This relationship is shown in Figure 7.8.

The value of the Total-Risk of a project will depend on the individual values of each Risk-Factor. By following this concept, the Total-Risk value will depend directly of the Risk-Factors providing this situation more real values of the risk in a project. The main point of Figure 7.8 is to show the "bond" between the Risk-Factors and the Total-Risk value.

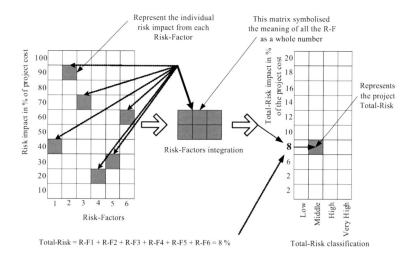

Figure 7.8   Risk-Factors and Total-Risk relation

The Risk-Factors and Total-Risk relation in Figure 7.8 works as follows. On the left side of the figure, there are six Risk-Factors. For each Risk-factor, a risk impact in % is given. For example, R-F1 (40 %), R-F2 (90 %), R-F3 (70 %), R-F4 (20 %), R-F5 (30 %) and R-F6 (60 %). Each value in %, represents the individual risk impact from each Risk-Factor.

The right side of Figure 7.8 represents the project Total-Risk, which is the integration of the six Risk-Factors. In this case, the impact of the Total-Risk is 8 %. The classification of the Total-Risk is illustrated, in this case only with four categories: low, middle, high and very high.

Figure 7.8 is divided into 3 parts: Risk-Factors (left side), Risk-Factors Integration (central part) and the Total-Risk Classification (right side). The relation with the NRAS is as follows:

Risk-Factors: Represents the input data of the system

- **Risk-Factors Integration**: Represents the internal structure of the system (hidden layers, algorithm, etc.)
- **Total-Risk**: Represents the output data of the system

By referring to the proposition made in Figure 7.1 due to the assessment of the Risk and Profit (W+G), as the basis of total production costs (HSK). The next step is to propose a way for assessing the Risk (W) as an independent factor. The above formula is proposed for evaluating the Total-Risk value (W_{T-R}).

(7.2) $$W_{T-R} = \frac{P_{current}}{HSK_{TPC}} \cdot 100$$

Where:

W_{T-R} = Total-Risk value

$P_{current}$ = Current profit

HSK_{TPC} = Total production costs

The justification of terms used in Formula 7.2 is based on the following statement. There are two variables acting in a project which indicate sensible values referring to the outcome of a project; these are the current profit ($P_{current}$) and the total production costs (HSK_{TPC}).

The relation between these two variables, produces a result (in this case W_{T-R}) which measures the Total-Risk value. Why Total-Risk? The answer is because the formula includes Total production costs and the final total profit acting in the whole project. The next question, why Risk? Is basically because the variable $P_{current}$ includes risks, for the reason that is common between contractors to use risk-premiums (risk fee) to protect their profit.

In addition, a similar situation happens with the variable HSK_{TPC}. In each production system or process, there are several risks involved. In the case of a contractor, the construction process is indeed a production process (the final product is finishing the project) which enclose along all the life cycle the participation of many risks.

As mentioned before, the Total-Risk (W_{T-R}) is the element in the NRAS which represents the outputs. This value is essential for developing further the system. By using a data example from a project (obtained from the Dresdner Company), together with formula 7.2, the Total-Risk value is calculated:

$$W_{T-R} = \frac{145{,}100.\text{-} \ \text{€}}{2{,}913{,}100.\text{-} \ \text{€}} \cdot 100$$

$W_{T-R} = 4.98\%$ (0.0498 in ANN values, see Figure 7.5)

The value for this example represents the output to be integrated into the artificial neural network together with the Risk-Factors.

Until now, no Risk-Factors have been used to calculate the Total-Risk value involved in a construction project. It is not possible to obtain the Risk-Factor values from a past project, only the total planned and actual costs are available. For that reason, and because this is the first time that Risk-Factors have taken into account the calculation of the Total-Risk together with artificial neural networks, the followings assumptions are made:

In order to clarify the use of the Risk-Factor Assessment framework (Table 7.3) a category and an impact was selected for each Risk-Factor, while using the data from project 5 (see table 8.1).

Risk-Factor Code	Risk-Factor	Risk-Factor Area	Risk-Factor Classification	Risk-Factor Score
R-F1	Risks caused by the change in law, regulations, etc.	Political	Middle	50
R-F2	Risks by problems on permissions	Political	E. Low	0
R-F3	Inflation risks	Economic	Very low	15
R-F4	Management risks caused by the subcontractor	Management	High	65
R-F5	Contract risks caused by the subcontractor	Management	High	65
R-F6	Risks from security and health protection	Management	E. Low	0
R-F7	Force majeure risks	Force Majeure	V. high	85
R-F8	Weather risks	Force Majeure	High	65
R-F9	Transport risks	Execution	High	65
R-F10	Design and construction risks	Execution	E. Low	0
R-F11	Quality risks	Execution	V. high	85
R-F12	Technical and execution risks	Execution	E. High	100
R-F13	Risks from water and air pollution	Execution	V. Low	15
R-F14	Contract risks	Contract	V. high	85
R-F15	Guarantee risks	Contract	V. high	85
R-F16	Business and Market risks	Contract	Middle	50
R-F17	Risks caused by the Client	Contract	Middle	50

Table 7.3   Risk-Factor Assessment framework example

The next step is to transform the Risk-Factor scores (Table 7.3, column 5) into neuronal values. This is done with the help of the Risk-Factors (inputs) ranking scale (Figure 7.4). For example for the R-F1: Risks caused by the change in law, regulations, etc; which has a Risk-Factor Score of 50 %. Then by checking on the middle column of Figure 7.4 and then taking its corresponding value in neuronal scores (third column, Figure 7.4), in this case equals to 0.50. The same procedure is follows for the remaining Risk-Factors.

7.8.3 Risk Evaluation Form and artificial neural network creation

The purpose of the Risk Evaluation Form (Figure 7.9) is to facilitate the control of the data related to the risks involved in the company's projects. Because of its simplicity, the Risk Evaluation Form will make it easier to maintain a risk-record per project. The Risk Evaluation Form consists of three parts: project data, risk factors and risk factor scale.

As can be seen, the Risk-Factors from Table 7.3, column 5 and the Total-Risk value from section 7.8.2, provide the data necessary to create the artificial neural network structure.

Risk Evaluation Form

Project Data

Date _____

Project _____ Offer amount _____

Project Manager _____ Construction Time _____

Client _____ Handing over Date _____

Construction Manager _____ User _____

RISK FACTORS

1	2	3	4	5
Risk Factor Code	Risk Factor	Risk Factor Area	Risk Factor Classification	Risk Factor Score
R-F1	Risks caused by the change in law, regulations, etc	Political		
R-F2	Risks by problems on permissions	Political		
R-F3	Inflation risks	Economic		
R-F4	Management risks caused by the subcontractor	Management		
R-F5	Contract risks caused by the subcontractor	Management		
R-F6	Risks from security and health protection	Management		
R-F7	Force majeure risks	Unforeseen		
R-F8	Weather risks	Unforeseen		
R-F9	Transport risks	Execution		
R-F10	Design and construction risks	Execution		
R-F11	Quality risks	Execution		
R-F12	Technical and execution risks	Execution		
R-F13	Risks from water and air pollution	Execution		
R-F14	Contract risks	Contract		
R-F15	Guarantee risks	Contract		
R-F16	Business and Market risks	Contract		
R-F17	Risks caused by the Client	Contract		

RISK FACTOR SCALE

Risk Factor Classification	Extreme High	Very High	High	Middle	Low	Very Low	Extreme Low
Risk Factor	100	85	65	50	35	15	0

Figure 7.9   Risk Evaluation Form

120 Theoretical Development of the Neuronal-Risk-Assessment System

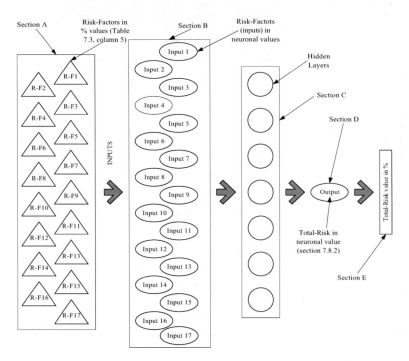

Figure 7.10   ANN structure

Figure 7.10 is divided into five sections:

Section A: Represents the 17 Risk-Factors which are listed in the Risk Evaluation Form. As is established in the Risk-Factors ranking scale (see Figure 7.4), the values for the Risk-Factors in this section are expressed as %.

Section B: Represents as in section before the 17 Risk-Factors, just the difference in this section is that the values of the Risk-Factors are expressed in neuronal values (see Figure 7.4). This section represents the input layer of the ANN structure.

Section C: Represents the hidden layer of the ANN structure. In this case, the number of neurons will be changeable as shown in Figure 8.5 (chapter 8) as part of the testing procedure for the NRAS.

Section D: Represents the value of the Total-Risk (calculated by Formula 7.2) in neuronal values (see Figure 7.5). This value represents the output (only one neuron in the output layer) of the ANN structure.

Section E: Represents the Total-Risk value (obtained in section D) in % (obtained while using the output ranking scale in Figure 7.5).

The data for training the artificial neural network is now complete (one training set in this case). However, this process will be carried on through the next chapter (beginning in sections 8.4, 8.5, 8.5.1, 8.5.2 and 8.5.3). The same steps will be followed for the complete set of data provided from other projects.

7.9   Neuronal-Risk-Assessment System data flow

The purpose of this section is to show how the complete set of data (of different projects) will be put together and then used for the creation of an ANN, which will be a new useful tool for the calculation of the Total-Risk.

The description of the "Data-Flow" (see Figure 7.11) is theoretical in this section. However the same procedure is adopted in chapter eight but at the application phase.

The structure of the Data-flow process is clear and logical. At this phase of the Neuronal-Risk-Assessment System procedure, the purpose is to create the ANN. The activities involved with the Data-Flow process are the ones described within the "Risk Management" (section 7.5.1) and the "Artificial intelligence" (section 7.5.2) components (see Figure 7.3).

After the ANN is created, it is then trained, and then the simulation is made and later on is the testing phase. That means that the ANN-Product is used together with data from new projects. Although this step (ANN-Product) is included in Figure 7.11.

For a better understanding, Figure 7.11 is divided in 8 sections, which are explained:

Data Source (section 1 and 8): This section encloses all the projects used to build up the NRAS. In other words, these are the data resources, which are in this case 16 projects.

Risk-Factors assessment (section 2): This section represents the complete set of Risk-Factors per project. In the actual work, 17 Risk-factors per 16 projects, giving a total of 272 Risk-Factors.

Risk-Factors values in % (section 3): This section deals with the values in % of each Risk-Factor per the 16 projects. As in the previous section, there are for the present work 272 Risk-Factors values in %.

Risk-Factors in Neuronal values (section 4): Represents exactly the same number (272) of Risk-Factors, but in this case in neuronal values, which were transformed by using the Risk-Factors (inputs) ranking scale (Figure 7.4), and as well, they form the input vector (see Figure 8.3, in chapter 8 in order to see the complete input vector for the 16 projects).

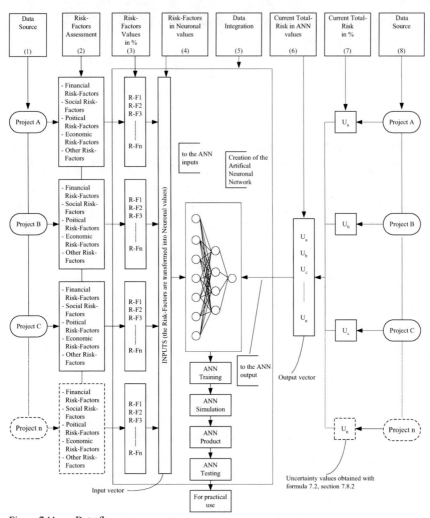

Figure 7.11   Data-flow process

<u>Data integration (section 5):</u> This section means putting together the input vector (section 4) and the output vector (section 6), in order to create the ANN structure (see Figure 7.10). After this, the elements: ANN training, ANN simulation, ANN product, ANN testing and the practical use are covered in sections 8.5 in chapter 8.

<u>Actual Project Total-Risk in ANN values (section 6):</u> Represents the value of the Total-Risk in neuronal values for each project (16 in this case). The 16 values of the Total-

Risk conforms the output vector. See Figure 8.4 from chapter 8, in order to see the complete output vector for the 16 projects.

Actual Project Total-Risk in % values (section 7): Encloses the data related to the values of the Total-Risk project (calculated with Formula 7.2) in % values. The % values are transformed into neuronal values (section 6) while using the Total-Risk (outputs) Ranking scale (Figure 7.5). Figure 7.12 describes in more detail how the ANN-Product will be used with new data. The activities involved in this phase are those described within the "Calculation of Construction Prices" (see Figure 7.3, section 7.5.3).

The purpose of Figure 7.12 is to show, how the NRAS will work with new projects, when the ANN is trained already. The following sections explain this process.

New project source of data (section 1): Represents the 17 Risk-Factors (in this work) which belong to a new project which has been not considered as part of the training data set of the ANN.

Risk-Factors assessment (section 2): Represents the scores assigned in the decision-making over the Risk-Factors (17 in this work). At this phase, the Risk-Factors are still in % values.

Risk-Factors transformation (section 3): Consist in transforming the % values of the Risk-Factors assessed in section 2, into neuronal values. This is done while using the Risk-Factors (inputs) ranking scale (Figure 7.4).

Use of the ANN-Product (section 4): Is in this section, where the ANN selected in the training and testing phases (this is done in sections 8.5.1 and 8.5.2 of chapter 8) is use to simulate the possible Total-Risk for a new project (done in section 8.5.3 of chapter 8)

Total-Risk value in neuronal terms (section 5): Represents the result obtained after the simulation carried out in section 4. The value of the Total-Risk at this section is in neuronal terms. This process is done in section 8.5.4 of chapter 8.

Total-Risk value in % (section 6): Consist of transforming the Total-Risk value (in neuronal terms obtained in section 5) into % values. This is done while using the Uncertainty (outputs) ranking scale (Figure 7.5). This is carried out in sections 8.5.4 and 8.5.5 of chapter 8.

Adding Total-Risk value into the Project costs (section 7): This section mean, that once the Total-Risk value is known in %. Then, this value is integrated to the project cost (as a % of the project cost).

It is important to mention, that all the data used in Figure 7.12 are data that represent a number of projects already completed. These data are the core element of the system in order to create the neural network. All the data will provide the system with the proposed Risk-Factors and the actual Total-Risk values from each finished project.

The main difference between Figure 7.11 and Figure 7.12, is that Figure 7.12 shows how the data will flow once the NRAS is ready in use with new projects. In other words, the figure shows how the NRAS will be feed with new Risk-Factors and the forecasting of the corresponding Total-Risk value. The other difference is that in Figure 7.12, it is not necessary to use the Total-Risk values from the finished projects because at this phase the NRAS is ready to be used in the practical world.

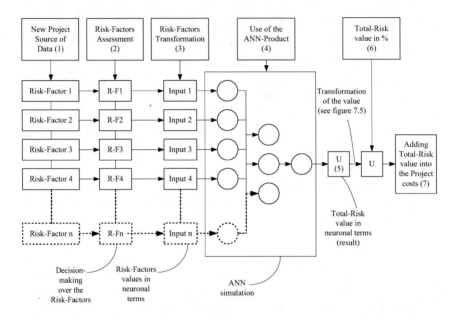

Figure 7.12   Use process of the Neuronal-Assessment System with new projects

7.10 Inputs and outputs model representation

The inputs and outputs model representation describes the link between the Risk-Factors and the Total-Risk value in mathematical terms. In order to clarify the Input-Output relationship used within the Neuronal-Risk-Assessment System, the following graphical and mathematical description is included:

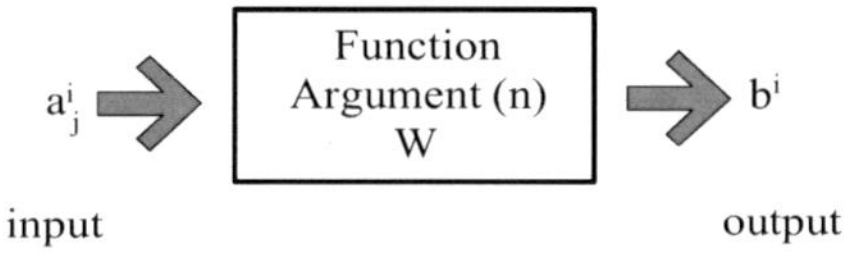

Theoretical Development of the Neuronal-Risk-Assessment System   125

$$a^i_j \rightarrow \boxed{\begin{array}{c}\text{Function}\\\text{Argument (n)}\\\text{W}\end{array}} \rightarrow b^i$$

input   output

Figure 7.13   Input-Output relationship

Where:

a   = is the input vector, where i represents the project number and j represents the Risk-Factors number; a is the value of the Risk-Factor
b   = is the output vector, where i represents the project number and b represents the value of the actual Total-Risk from each project
n   = is the argument of the transfer function f (function argument)
W   = are the scalar weights

The Input and Output representation as vectors are as follows:

$$\begin{pmatrix} a^1_1 & a^2_1 & \cdots & a^n_n; \\ a^1_2 & a^2_2 & \cdots & a^n_n; \\ \vdots & \vdots & \vdots & \vdots \\ a^n_n & a^n_n & \cdots & a^n_n; \end{pmatrix} \quad \begin{bmatrix} b^1 & b^2 & b^3 & b^4 & \cdots & b^n \end{bmatrix}$$

                  Input Vector   Output Vector

Figure 7.14   ANN vectors

The mathematical model of the neuron, which represents the Input-Output relation in this research work, is as follows:

$$X = (X_1, X_2, \ldots, X_{17})$$
$$W = (W_1, W_2, \ldots, W_{17}), b$$

(7.3) $\quad Y = f\left(\sum_j W_j X_j - b\right); \quad n = \left(\sum_j W_j X_j - b\right)$

$$Y = f(n)$$

Where:

Y = is the output (denotes the Total-Risk value)
f = is the chosen transfer function
X = are the inputs (in this case, represent the input values of the Risk-Factors)
W = are the scalar weights
b = bias
n = argument of the transfer function
j = represents the Risk-Factors numbers

The example carried out in this chapter was to show in greater detail how the system is formulated, in this case by only using the data (Risk-Factors and Total-Risk value) of only one project. In chapter eight, the complete set of data (16 projects) will be used for developing a practical example of the NRAS. The purpose of chapter 7 was to explain each step of the system.

7.11 Analysis of the expected results

The expected behaviour of the NRAS is unknown at this stage. It is not possible before hand to know the possible results that an ANN will produce.

However, the desire is that the NRAS will be able to imitate the actual results with a minimum error and give acceptable results. Figure 7.15 shows the possible expected results of the NRAS (ANN Total-Risk values), together with the actual Total-Risk values and the regularly Total-Risk values in literature.

As was mentioned at the beginning of this chapter, at the present there is no systematic approach available to be used in calculating the risk cost. For that reason the literature regularly proposes a risk value of 2 % (sometimes 3 %). This is independent of the project's uniqueness, complexity, and place of development. In other words, without taking into account the possible impacts of the project's risk factors with due regard the risk value for each project.

It is concluded that the value assigned to the risk in construction projects does not behave like in reality. It can be straightforward to decide the risk value, but as well, it is very easy indeed to fall in catastrophically loses within the project and well thorough the company. It can be fairly argued by contractors and researchers that the number of Risk-Factors used on this work is limited and that only resembles the risk environment of a German construction company in a specified developed country. Nevertheless, to solve the problems and "head aches" that the risks cause in the construction industry, cannot be solve with a single recipe or formula. To properly solve the impact of risks in construction companies, demands a constant development of new tools to be adapted and applied in specific areas of construction and parts of the world. Risk management demands a serious and formal compromise not only from the top management levels of the company; demands as well a strong commitment from all the operative resources which in one point will collaborate to feed back data relevant for the management of risks. The goal searched with the Neuronal-Risk-Assessment System is that the Total-Risk will be unique for each project. That means that this value will behave in accordance with the assessed impact of the Risk-Factors in each project (see Figure 7.15). The main advantage of the Neuronal-Risk-Assessment System is that the value of the Total-Risk will correspond to actual project risks.

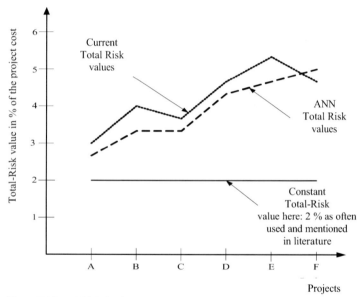

Figure 7.15   Risk behaviour

Further, it is possible to use the concept and value of the Risk and Profit (W+G) in accordance to the ARRIBA construction software (see Figure 7.16). This figure is used with the purpose of showing the options available for selecting the Risk + Profit values within ARRIBA construction software.

As can be seen in Figure 7.16 the Risk + Profit values can be selected as constant values or can be proposed. The Risk and Profit are considered together, which in the case of the NRAS proposed that the Total-Risk must be calculated independent to the profit (G). This is a good opportunity to use the results obtained with the Neuronal-Risk-Assessment System.

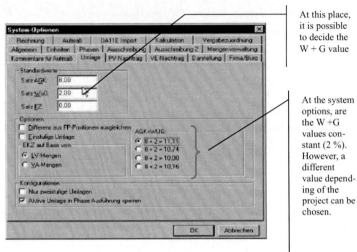

At this place, it is possible to decide the W + G value

At the system options, are the W +G values constant (2 %). However, a different value depending of the project can be chosen.

Figure 7.16   Risk Value with the ARRIBA software

8 Empirical Implementation of the Neuronal-Risk-Assessment System (NRAS)

8.1 Introduction

In this application example of the NRAS, a set of data of 16 real projects is used for testing the usefulness and reliability of the system. In section 8.2, the assessment of the Risk-Factors (ANN inputs) for each of the 16 projects is explained when using the Risk Evaluation Form (Figure 7.9). This is done because from this data, the inputs necessary for training and testing (see section 8.4) the ANN are obtained. Each Risk-Factor will be evaluated and classified and then, the Risk-Factors will be scored. With these score values (see Risk Evaluation Form, Figure 7.9 and Risk-Factors column 5), finally they will be transformed into neuronal values with the help of the Risk-Factors (inputs) ranking scale (see Figure 7.4). In order to see the complete set of input data, refer to section 8.4, (Figure 8.3). In addition, an argument is made in relation to the meaning of the Risk-Factors as risk profiles.

In section 8.3, the "Total-Risk" (ANN outputs) for each project will be calculated. The procedure used to appraise the Total-Risk value was indicated in section 7.8.2. After the Total-Risk values (in %) are obtained for each project, the next step is to transform these values into ANN terms (as explained in section 7.8.2). This is done with the help of the Total-Risk (output) ranking scale (see Figure 7.5). In order to see the complete set of output data, refer to section 8.4, (Figure 8.4). Section 8.4, deals with the design of the ANN structure. For that purpose, using the data from 12 projects; the input vector (Figure 8.3) and output vector (Figure 8.4) are created. Section 8.5 deals with the process of testing the NRAS. It includes: testing procedure of the NRAS (section 8.5.1) For a better understanding, Figure 8.5 explains in detail the procedure adopted to probe the NRAS, using 12 projects for training the ANN and the remaining 4 projects for validating the performance of the NRAS.

Backpropagation network performance at the training phase (section 8.5.2), based on the testing procedure (Figure 8.5), a number of ten NRAS models were created (from NRAS(A) to NRAS(J)), and with this the creation and training of a total 100 ANNs. NRAS results (section 8.5.3), in this section, the results for each of the ten ANNs (BP1 to BP10) created in the NRAS(A) model are shown, these results are included in Figure 8.7. A discussion is made towards the results offered by the NRAS(A) model. The results obtained with model NRAS(C) (Figure 8.8) are shown for complementing this section.

The meaning and impact of the NRAS results are shown in section 8.5.4. In addition, a description of the meaning of the results of the NRAS(A) model is included, as a comparison with the actual values proposed without the use of the NRAS. This comparison is done in % values and in money terms. The intention of this section (section 8.5.5) is to evaluated how well was each Risk-Factor assessed; with the actual approach and the NRAS. Taking into account the classification and the score of the Risk-Factors.

8.2 Risk-Factors assessment

This section provides the input data for the creation of the ANNs, which are developed in section 8.5. A set of 16 finished projects from a Dresdner construction company have been used.

All these data are enclosed within the Risk-Factors scores (Table 8.1). The data represent the decisions made over the Risk-Factors for each project. The decisions were made by a project manager who was working closely with the projects and they were based on their consequence and probability.

The results of each Risk-Profile are put together into an input vector and then transformed into ANN values. See section 8.4

8.2.1 Meaning of the Risk-Factors as risk profiles

Each project was evaluated by an expert (in this case the estimator and the branch manager) from the risk management point of view. That means that the possible risks (Risk-Factors) of each project were assessed in order to build up a risk profile, which will be the basis for obtaining the uncertainty value in future projects. Therefore it will be possible to know the Total-Risk values for new projects by just making a decision on the Risk-Factors.

The Risk-Factors scores for each project are enclosed in Table 8.1. The risk profiles of two projects are included to illustrate how the Risk-Factors vary between projects.

Empirical Implementation of the Neuronal-Risk-Assessment System 131

Risk-Factors

Project	RF-1	RF-2	RF-3	RF-4	RF-5	RF-6	RF-7	RF-8	RF-9	RF-10	RF-11	RF-12	RF-13	RF-14	RF-15	RF-16	RF-17
1	50	15	15	85	85	0	65	65	65	0	85	100	0	65	65	15	50
2	15	0	0	85	85	0	65	85	100	0	85	100	0	100	85	15	50
3	50	15	15	85	65	15	85	65	100	15	85	100	15	85	85	50	35
4	35	0	15	65	85	15	100	85	100	0	85	100	15	35	50	50	35
5	50	0	15	65	65	0	85	65	65	0	85	100	15	85	85	50	50
6	35	15	15	50	50	15	65	65	65	15	65	100	15	65	65	35	65
7	35	0	15	85	85	0	65	85	100	0	100	100	0	85	65	35	50
8	35	15	0	50	50	15	65	65	85	15	65	100	0	50	50	15	65
9	35	0	0	50	65	15	65	85	85	0	65	100	15	50	50	35	35
10	50	15	35	65	85	65	50	35	85	15	35	85	100	85	35	50	85
11	35	15	35	85	85	0	65	85	100	15	65	85	15	35	65	35	50
12	15	0	0	50	50	15	85	85	65	15	100	100	15	65	65	50	50
13	15	35	35	50	50	15	65	85	15	35	50	85	15	50	35	15	0
14	50	15	0	50	50	0	35	35	50	0	35	100	0	50	50	35	50
15	35	0	0	50	65	0	65	65	85	0	65	100	15	50	50	50	50
16	15	0	0	35	50	65	50	85	65	15	0	100	15	50	0	0	35

Table 8.1   Risk-Factors scores

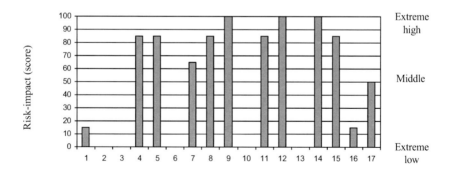

Figure 8.1   Risk-Profile for project 2

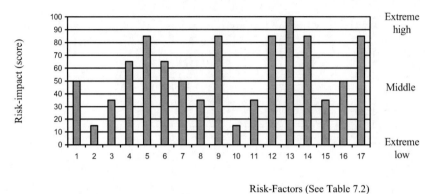

Figure 8.2   Risk-Profile for project 10

Figure 8.1 shows the risk impacts for project 2. It is clear to notice that the Risk-Factors: 4, 5, 8, 9, 11, 12, 14 and 15 were evaluated by the experts with the highest scores. That means that for this company at the time at which the project was undertaken the critical Risk-Factors were: *Management risks caused by the subcontractor, contract risks caused by the subcontractor, weather risks, transport risks, quality risks and technical and execution risks.*

There were other Risk-Factors which were classified based on their scores as not so dangerous. These were the Risk-Factors 7 and 17: *Force majeure risks and risks caused by the client.* As well, there were two Risk-Factors which were classified as moderate; these were the Risk-Factors 1 and 16, which are: *Risks caused by the change in law, regulations, etc; and Business and market risks.*

Nevertheless, there were some Risk-Factors which were considered by the experts as extremely low based on their scores. These Risk-Factors were: 2, 3, 6, 10 and 13; which are: *Risks by problems of permissions, inflation risks, risks from security and health protection, design and construction risks and risks from water and air pollution.*

With the other Risk-Profile for project 10 (Figure 8.2) the situation is other. In this case, the experts had evaluated as the Risk-Factors 5, 9, 12, 13, 14 and 17 as the critical ones. Which are: *Contract risks caused by the subcontractor, transport risks, technical and execution risks, risks from water and air pollution, contract risks and risks caused by the client.*

This is different to the Risk-Profile of project 2 (Figure 8.1). In this case, some Risk-Factors were evaluated as to be of middle impact. These were: 1, 4, 6 and 16. Which are: *Risks caused by the change in law, regulations, etc; management risks caused by the subcontractor, risks from security and health protection and business and market risks.*

There was no Risk-Factor classified as extremely low in this case, but the rest were classified as moderate. These are 3, 7, 8, 11 and 15; which are: *Inflation risks, force majeure risks, weather risks, quality risks and guarantee risks.*

Finally, the rest of the Risk-Factors were classified as very low. These were 2 and 10; which are: *Risks by problems on permissions and design and construction risks.*

It is important to notice that in this case for this company; Risk-Factors like 5, 9, 12, and 14; were classified in both Risk-Profiles (Figures 8.1 and 8.2) as very high or extremely high. This situation shows the awareness of the company towards specific Risk-Factors and also shows that the company is more orientated towards operative and control tasks.

From all the Risk-Factors the ones corresponding to the technical and execution area/phase seriously affected the company's financial success. That means that this company is more orientated towards operation and control tasks. The reason is because the company branch from which the data was taken, is mainly focussed at the operational management tasks; with a small concentration at the middle management tasks.

8.3 Total-Risk assessment

Tables 8.2 and 8.3 show the data from the 16 projects employed. The data is divided into planned costs (Table 8.2) and actual costs (Table 8.3).

With the data from Table 8.3, together with formula 7.2, the value of the current Total-Risk for each project is calculated (See Table 8.3, column 7). After this, the complete set of data is transformed into ANN values while using the Total-Risk (output) ranking scale (see Figure 7.5). Finally, the values of the current Total-Risk (W_{T-R} , Table 8.7, column 7) are show as the output vector in section 8.4.

Project	Production Costs (HSK)	General Business Costs (AGK)	General Business Costs (AGK)	(W+G) planned (with 2 % of AS)	Net Offer Amount (AS netto)
Nr	€	€	in % of AS	€	€
1	2	3	$4 = \dfrac{3}{6}$	$5 = 0.02 \cdot 6$	$6 = 2 + 3 + 5$
1	832,038.57	75,976.77	8.20	18,530.93	926,546.27
2	843,899.35	122,303.59	12.40	19,718.43	985,921.37
3	471,010.44	54,330.43	10.13	10,721.24	536,062.11
4	4,833,199.61	473,887.25	8.75	108,307.90	5,415,394.76
5	2,018,564.07	275,296.88	11.76	46,813.49	2,340,674.44
6	1,267,735.92	219,231.04	14.45	30,346.26	1,517,313.22
7	1,659,940.73	280,523.04	14.17	39,601.30	1,980,065.07
8	1,854,519.56	297,345.14	13.54	43,915.61	2,195,780.31
9	503,222.67	21,363.91	4.00	10,705.85	535,292.43
10	2,517,417.16	161,680.75	5.91	54,675.47	2,733,773.38
11	438,511.49	45,978.81	9.30	9,887.57	494,377.87
12	435,144.01	65,562.68	12.83	10,218.50	510,925.19
13	1,448,906.33	233,686.03	13.61	34,338.62	1,716,930.98
14	3,845,087.12	274,161.78	6.52	84,066.30	4,203,315.20
15	1,316,481.35	114,419.26	7.84	29,202.05	1,460,102.66
16	478,512.45	67,086.34	12.05	11,134.67	556,733.46

Table 8.2   Planned costs

It is important to notice in column 5 of Table 8.2, that the value W+G does not change. That means it is considered as 2 % for all projects. This value (W+G, 2 %) is obtained from the AS netto (Column 6, Table 8.2) from each project.

Project	Performance	Total Production Costs (HSK_{TPC})	Total Profit	General Business Costs (AGK)	Current Profit ($P_{current}$)	Total-Risk (W_{T-R})
Nr	€	€	€	€	€	in % of Production Costs
1	2	3	4 = 2 - 3	5	6 = 4 - 5	$7 = \dfrac{6}{3}$
1	1,825,100.00	1,285,900.00	539,200.00	179,300.00	359,900.00	27.99
2	2,440,300.00	2,164,600.00	275,700.00	286,000.00	-10,300.00	-0.47
3	3,584,800.00	3,134,400.00	450,400.00	395,600.00	54,800.00	1.75
4	6,459,700.00	5,855,900.00	603,800.00	694,800.00	-91,000.00	-1.55
5	3,105,700.00	2,610,600.00	495,100.00	346,400.00	148,700.00	5.69
6	3,443,800.00	2,913,100.00	530,700.00	385,600.00	145,100.00	4.98
7	2,250,300.00	1,936,700.00	313,600.00	256,800.00	56,800.00	2.93
8	2,669,500.00	3,112,300.00	-442,800.00	248,500.00	-691,300.00	-22.21
9	206,600.00	227,700.00	-21,100.00	18,400.00	-39,500.00	-17.35
10	3,206,800.00	3,042,300.00	164,500.00	142,500.00	22,000.00	0.72
11	739,600.00	849,600.00	-110,900.00	85,100.00	-195,100.00	-22.96
12	585,900.00	553,900.00	32,000.00	58,200.00	-26,200.00	-4.73
13	2,031,800.00	1,842,200.00	189,600.00	197,300.00	-7,700.00	-0.42
14	3,828,800.00	3,562,600.00	266,200.00	438,400.00	-172,200.00	-4.83
15	1,410,400.00	1,245,200.00	165,200.00	153,800.00	11,400.00	0.92
16	698,900.00	561,400.00	137,500.00	26,100.00	111,400.00	19.84

Table 8.3   Actual costs

In this case with the actual results, the value of the current Total-Risk (W_{T-R}, column7, Table 8.3) is different for each project. In some cases, it is even negative. These values show the reality of the Total-Risk values in construction projects, are unique per each project and are not constant (for example 2 %) as assumed in Table 8.2, column 5.

8.4   ANN structure

Before starting with the training and the simulation of the ANN, the inputs and the outputs are arranged as vectors (input and output vectors) as indicated in section 7.8.3 (Figure 7.10). It is necessary to organise the data in order to give the ANN the freedom to read and use them. The input (from Table 8.1) and output (from Table 8.3, column 7) vectors are as follows:

$$\begin{pmatrix}
0.50 & 0.15 & 0.50 & 0.35 & 0.50 & 0.35 & 0.35 & 0.35 & 0.35 & 0.50 & 0.35 & 0.15 & 0.15 & 0.50 & 0.35 & 0.15 \\
0.15 & 0 & 0.15 & 0 & 0 & 0.15 & 0 & 0.15 & 0 & 0.15 & 0.15 & 0 & 0.35 & 0.15 & 0 & 0 \\
0.15 & 0 & 0.15 & 0.15 & 0.15 & 0.15 & 0.15 & 0 & 0 & 0.35 & 0.35 & 0 & 0.35 & 0 & 0 & 0 \\
0.85 & 0.85 & 0.85 & 0.65 & 0.65 & 0.50 & 0.85 & 0.50 & 0.50 & 0.65 & 0.85 & 0.50 & 0.50 & 0.50 & 0.50 & 0.35 \\
0.85 & 0.85 & 0.65 & 0.85 & 0.65 & 0.50 & 0.85 & 0.50 & 0.65 & 0.85 & 0.85 & 0.50 & 0.50 & 0.50 & 0.65 & 0.50 \\
0 & 0 & 0.15 & 0.15 & 0 & 0.15 & 0 & 0.15 & 0.15 & 0.65 & 0 & 0.15 & 0.15 & 0 & 0 & 0.65 \\
0.65 & 0.65 & 0.85 & 1 & 0.85 & 0.65 & 0.65 & 0.65 & 0.65 & 0.50 & 0.65 & 0.85 & 0.65 & 0.65 & 0.35 & 0.50 \\
0.65 & 0.85 & 0.65 & 0.85 & 0.65 & 0.65 & 0.85 & 0.65 & 0.85 & 0.35 & 0.85 & 0.85 & 0.85 & 0.35 & 0.65 & 0.85 \\
0.65 & 1 & 1 & 1 & 0.65 & 0.65 & 1 & 0.85 & 0.85 & 0.85 & 1 & 0.65 & 0.15 & 0.50 & 0.85 & 0.65 \\
0 & 0 & 0.15 & 0 & 0 & 0.15 & 0 & 0.15 & 0 & 0.15 & 0.15 & 0.15 & 0.35 & 0 & 0 & 0.65 \\
0.85 & 0.85 & 0.85 & 0.85 & 0.85 & 0.65 & 1 & 0.65 & 0.65 & 0.35 & 0.65 & 1 & 0.50 & 0.35 & 0.65 & 0 \\
1 & 0.85 & 0.85 & 0.85 & 0.85 & 0.65 & 1 & 1 & 1 & 0.85 & 0.85 & 1 & 0.85 & 1 & 1 & 1 \\
0 & 0 & 0.15 & 0.15 & 0.15 & 0.15 & 0 & 0 & 0.15 & 1 & 0.15 & 0.15 & 0.15 & 0 & 0.15 & 0.15 \\
0.65 & 1 & 0.85 & 0.35 & 0.85 & 0.65 & 0.85 & 0.50 & 0.50 & 0.85 & 0.35 & 0.65 & 0.50 & 0.50 & 0.50 & 0.50 \\
0.65 & 0.85 & 0.85 & 0.50 & 0.85 & 0.65 & 0.65 & 0.50 & 0.50 & 0.35 & 0.65 & 0.65 & 0.35 & 0.50 & 0.50 & 0 \\
0.15 & 0.15 & 0.50 & 0.50 & 0.50 & 0.35 & 0.35 & 0.15 & 0.35 & 0.50 & 0.35 & 0.50 & 0.15 & 0.35 & 0.50 & 0 \\
0.50 & 0.50 & 0.35 & 0.35 & 0.50 & 0.65 & 0.50 & 0.65 & 0.35 & 0.85 & 0.50 & 0.50 & 0 & 0.50 & 0.50 & 0.35
\end{pmatrix}$$

The columns represent the Risk-Factors per project, see Figures 7.13 and 7.14

The elements along the rows represent the 17 Risk-Factors, see Figures 7.13 and 7.14

Figure 8.3   Input vector

$$\begin{pmatrix} 0.2799 & 0.0047 & 0.0175 & 0.0155 & 0.0569 & 0.0498 & 0.0293 & 0.2221 & 0.1735 & 0.0072 & 0.2296 & 0.0473 & 0.0042 & 0.0483 & 0.0092 & 0.1984 \end{pmatrix}$$

These values are the Total-Risk for each project (see Figures 7.13 and 7.14)

Figure 8.4   Output vector

The values for both, the inputs and outputs are transformed into neuronal values with the help of Figures 7.4 and 7.5 (in chapter 7). With this data, it is now possible to train and simulate different structures of ANNs. This is done in the next section.

8.5 Testing the NRAS

8.5.1 Testing procedure of the NRAS

The procedure for testing the NRAS is described in Figure 8.5. The purpose of this procedure is to create different scenarios of the NRAS (10 different models) for evaluating the system.

From the complete set of data of 16 projects, 12 projects (from project 1 to project12) were chosen for being the data for training the system and the remaining 4 projects (from project 13 to project 16) were used for testing the system. The reason of this choice is basically random, assuming that each project has the same number of Risk-Factors and the data is limited to 16 finished projects.

With the 12 projects, 10 different models of the NRAS were created. The difference of each model is the order chosen for each project for training the system. For example, in the case of the NRAS(A) model (see Figure 8.5), the order chosen for the 12 projects was 1, 2, 3, 4, 5, 6, 7, 8, 9, 10, 11, 12. In the other case for example with the model NRAS(D) (see Figure 8.5) the order chosen was 2, 4, 6, 8, 10, 12, 1, 2, 5, 7, 9, 11. This criterion was used to create the NRAS models for training and then evaluating the system.

As well for each model, 10 different ANNs were created. The difference of each ANN is the number of neurons in the hidden layer. For example, BP1 (see Figure 8.5) means that is a backpropagation neural network with one neuron in the hidden layer. A BP7 network (see Figure 8.5) means that is a backpropagation neural network with 7 neurons in the hidden layer. This was done for each model, creating a total of 100 ANNs.

At the testing phase (see Figure 8.5), the remaining 4 projects were used for evaluating the system. These projects were: projects 13, 14, 15 and 16. For each ANN (BP1 to BP10) from each model, its mean squared error (MSE) was obtained in order to measure the performance of each network.

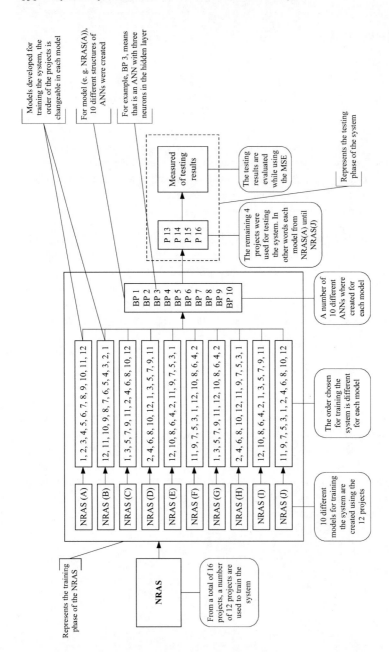

Figure 8.5   NRAS testing procedure

8.5.2 Backpropagation network performance at the training phase

A set of neural networks (10 for each model) were trained and their errors towards the whole set of outputs measured. This is shown in Figure 8.6, which includes the example for the NRAS(A). The examples for the rest of the models are shown in Appendix II. All the networks were trained with the same parameters: 1000 epochs, LMS learning algorithm and the log-sigmoid transfer function.

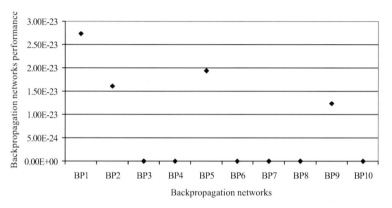

Figure 8.6   Backpropagation network performance for NRAS(A)

The information that provides Figure 8.6, is related to the MSE values obtained while doing the performance at the backpropagation algorithm with 10 networks of the model NRAS(A). The error values, in this case were obtained with the training process; nevertheless, the operational way for calculating these values is similar to the one used in chapter 5 (section 5.7.1). The meaning of these error values is how well at the training phase the network imitates all the output values (desired output values).

As can be seen in Figure 8.6, the errors obtained at the training phase for each network (BP) were negligible. In some cases, for example with the networks: BP1, BP2, BP5 and BP9, the errors were bigger than for the rest of the networks (BP3, BP4, BP6, BP7, BP8 and BP10) trained for the NRAS(A) model. In all cases, the errors are not significant for a good performance of the networks at this phase. The reason of why in the cases of networks BP3, BP4, BP6, BP7, BP8 and BP10 (Figure 8.6) the values are cero, is because the values obtained with the simulation at the training phase by these networks were the same as the planned values.

In the NRAS(C), NRAS(E), NRAS(H), NRAS(I) and NRAS(J) models; almost all the networks behave as well with very minor errors, just some networks present a higher error. For example in model NSAS(C) (see Appendix II) was the network BP10 (which presents errors as: P13 = 0.0251, P14 = 0.000000337, P15 = 0.00138 and P16 = 0.0915) , in model NRAS(E) (see Appendix II) was the network BP9 (which presents errors as: P13 = 0.000713, P14 = 0.0152, P15 = 0.0000418 and P16 = 0.00921), in model NRAS(H) (see Appendix II) was the network BP1 (which presents errors as: P13 = 0.00690, P14 = 0.00738, P15 = 0.2084 and P16 = 0.000553), in model NRAS(I) (see Appendix II) was the network BP7 (which presents errors as: P13 = 0.000856, P14 = 0.000747, P15 = 0.0103 and P16 = 0.00000261) and in model NRAS (J) (see Appendix II) was the network BP5 (which presents errors as: P13 = 0.0346, P14= 0.00187, P15 = 0.0000634 and P16 = 0.0907). For reviewing the behavior of the networks for the rest of the models, refer to Appendix II. The meaning of all these error values, is that show how well each network imitated the desired output. In other words, each network was trained with a desired output value and then simulated to obtain a "simulated output value", and the difference between these output values for each chosen project represents the error.

It must be recognized, that in general the performance of the ANNs for each of the 10 models was very acceptable (because the MSE errors were relative small at the training phase). Nevertheless, it is important to mention that perhaps the reason of this behavior of mainly all the networks at this training phase was due to the relative small data set available. However, in the ANN world, it is unpredictable before hand to know if a neural network will be a good solution for the problem intended to be solved or that the neural network will produce acceptable results.

8.5.3 NRAS results

For each of the 10 NRAS models (see Figure 8.5), 12 projects were used to train each NRAS model and 4 projects (13, 14, 15 ,16) were used to test each model through the simulation. An example of the testing results in this case for models NRAS(A) and NRAS(C) are described in Figures 8.7 and 8.8.

It was known before that the data available for training the ANN and for doing the test of the ANN was minimal and represented a critical point. This situation was discussed at the Neuronal Research Group of the Technische Universität Dresden, which is headed by Dr. Flach. The conclusion obtained with this group was that even do that the data size was very small, it was enough to train and test the system.

The Mean Squared Error (MSE) was used to evaluate the ANN's performance. This is define as:

(8.1) $$MSE = \frac{1}{n}\sum_{i=1}^{n}(X_1 - X_2)^2$$

Where:
n = total number of data points used for the model
X_1 = is the desired output (target)
X_2 = is the output obtained with the simulation

For example, for determining the MSE value for project 13 in NRAS(A) for BP1:

$$MSE = \frac{1}{12}(-0.0042 - 0.2817)^2 = 0.00681$$

The MSE results for the testing set are shown in Figure 8.7 for each ANN developed along the NRAS(A) model. In some networks at the testing phase, the MSE values were higher. For example in the BP1 (project 1), BP5 (project 16), BP7 (project 16), BP8 (project 16), BP9 (project 16) and BP 10(project 16). With the rest of the networks the MSE values were smaller.

In general, the errors were minor, taking into account that the whole set of data available for training and testing is very limited. The MSE for the rest of the networks developed with the other models (NRAS(B) to NRAS(J)) are enclosed in Appendix II.

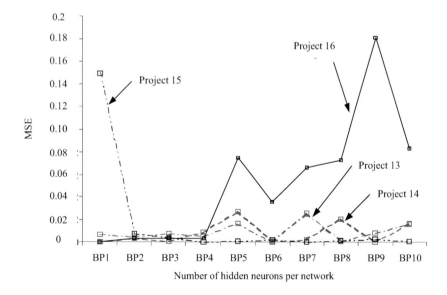

Figure 8.7   NRAS(A) results

Figure 8.7 enclosed the results in terms of the MSE obtained at the testing phase. In this case, the values correspond to the NRAS(A) model. The above figure shows the MSE for ten different networks (from BP1 to BP10) simulated with the testing data (projects 13, 14, 15 and 16). It is important to remember that BP1 means that the network has only one neuron at the hidden layer.

It is possible to observe (from Figure 8.7) that for some projects in all networks, the MSE was very small. For example, this situation happen with network BP1 (project 13, 14 and 16), BP2 (project 13, 14, 15 and 16), BP3 (project 13, 14, 15 and 16), BP4 (project 13, 14, 15 and 16), BP5 (project 13, 14 and 15), BP6 (project 13, 14 and 15), BP7 (project 13, 14 and 15), BP8 (project 13, 14 and 15), BP9 (project 13, 14 and 15) and BP10 (project 13, 14 and 15).

There are some MSE values which are considerably bigger in comparison with the ones described at the above paragraph. This situation happens at network BP1 (project 15), BP5 (project 16), BP8 (project 16), BP9 (project 16) and BP10 (project 16), but only in one project per network. The results obtained with the NRAS(C) model are included in Figure 8.8, in order to see the behaviour of the MSE in a different model of the NRAS.

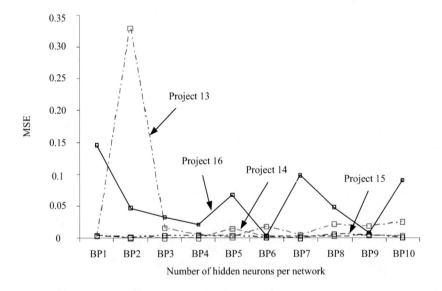

Figure 8.8   NRAS(C) results

The behaviour of the NRAS(C) model is different to the one of the NRAS(A) model. The NRAS(C) presents for example that the highest MSE error is in network BP2 (project 13). At the NRAS(A) model, the highest MSE value occurred at network BP9 (project 16). While continuing with the NRAS(C) model, very small MSE values were presented in network BP1(project 14 and 15), BP2 (project 14 and 15), BP3 (project 13, 14, 15 and 16), BP4 (project 13, 14, 15 and 16), BP5 (project 13, 14 and 15), BP6 (project 13, 14, 15 and 16), BP7 (project 13, 14 and 15), BP8 (project 13, 14 and 15), BP9 (project 13, 14, 15 and 16) and BP10 (project 13, 14 and 15). Theses results show the behaviour of the model (NRAS(C)) while simulating the data of projects 13 to 16 at the testing phase. In other words, these results show how accurate was the result (output = Total Risk W_{T-R}) for each project and the error is measured in terms of the MSE.

A comparable situation at this testing phase happened with the rest of the NRAS models from NRAS(B) to NRAS(J); see appendix II.

It is clear to see that at the testing phase the correspondence between the desired and actual outputs was not imitated as happens with the training phase. In order words, there was a difference between the desired and actual outputs when using the testing set of projects (P13, P14, P15 and P16). In the case for the NRAS(A) model and, indeed, for all the other models (from NRAS(B) to NRAS(J)), 4 projects were used for testing the system while observing if the system was able to recognize that these patterns (from these 4 projects) do not fit the pattern of the others.

An analogous situation happened for the 100 ANNs created for the NRAS models (see Appendix II) regarding the testing set. This behaviour of the networks towards the simulation of the actual output at the testing phase for all the models is attributed to the relative small numbers of projects available to train and to test the system.

However, the system was able to recognize that the patterns used with the 4 projects for testing the system (simulation phase) do not belong to the patterns used at the training phase. The system was able to recognize the patterns of the projects used at the training phase. Even so, in some cases the system was able to simulate at least 2 actual outputs closely to the desired outputs as happened with the network BP6 in Figure 8.7.

8.5.4 The meaning and impact of the NRAS results

Along this section, the results obtained with the NRAS at the testing phase are used in order to compare these results with the results from the actual procedure used for assessing the Total-Risk (W_{T-R}).

With the purpose for comparing the results obtained with the NRAS at the testing phase with the actual procedure use for assessing the Total-Risk (W_{T-R}). The network BP10 from model NRAS(E) was chosen (because its errors per project were minimal as shown in the above figure) to make this comparison. The BP10 errors versus outputs at the training phase are shown in Figure 8.9.

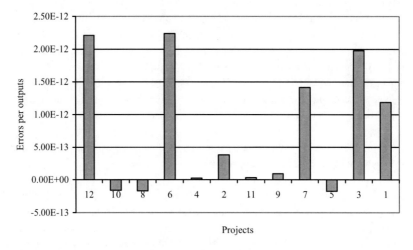

Figure 8.9   BP10 errors from model NRAS(E)

The BP10 produced small errors for projects 2, 4, 5, 8, 9, 10, 11. Most for the other projects are higher. Nevertheless, it can be mentioned, that in general the errors per output offered by the BP10 network are small and useful for the further analysis. The errors per output shown in Figure 8.9 were obtained by the simulation process. Although, the operational process to assess these values is the same as the one described in chapter 5 (section 5.7.1).

Tables 8.4 and 8.5, show the results related to the W_{T-R} values. The first table (Table 8.4), shows the results obtained with the BP10 from model NRAS(E) and Table 8.5 shows the results obtained with the actual process available.

	Project	Current Total-Risk in %	Total-Risk (with ANN) in %	Difference with the actual Total-Risk	MSE (Mean squared error)
Training phase	12	- 4.73	- 4.73	0	0
	10	0.72	0.72	0	0
	8	- 22.21	- 22.21	0	0
	6	4.98	4.98	0	0
	4	- 1.55	- 1.55	0	0
	2	- 0.47	- 0.47	0	0
	11	- 22.96	- 22.96	0	0
	9	- 17.35	- 17.35	0	0
	7	2.93	2.93	0	0
	5	5.69	5.69	0	0
	3	1.75	1.75	0	0
	1	27.99	27.99	0	0
Testing phase	13	- 0.42	- 4.10	3.68	0.000112
	14	- 4.83	- 8.35	3.52	0.000103
	15	0.92	- 21.90	22.82	0.00433
	16	19.84	8.65	11.19	0.00104

Table 8.4   Total-Risk with NRAS(E) using network BP10

As can be seen from Table 8.4, the network results for all the projects used for the training set are acceptable because the simulated Total-Risk (with ANN), imitates very or exactly well the current Total-Risk. There is practically no difference between the current Total-Risk and the "Total-Risk (with the ANN)". By consequence, the MSE value is zero from projects 1 to 12.

At the testing set, the reality is different because the network recognized that the patterns used for the testing set (projects 13, 14, 15, 16) do not fit to the patterns used at the training phase. In this case, the values of the "Total-Risk with (ANN)" present a difference to the actual current Total-Risk values (see Table 8.4, column 4).

146 Empirical Implementation of the Neuronal-Risk-Assessment System

	Project	Actual Total-Risk in %	Total-Risk (actual process) in %	Difference with the actual Total-Risk	MSE (Mean squared error)
Training phase	12	- 4.73	2.00	6.73	0.000377
	10	0.72	2.00	2.72	0.0000617
	8	- 22.21	2.00	24.21	0.00488
	6	4.98	2.00	6.98	0.000406
	4	- 1.55	2.00	3.55	0.000105
	2	- 0.47	2.00	2.47	0.0000508
	11	- 22.96	2.00	24.96	0.00519
	9	- 17.35	2.00	19.35	0.00312
	7	2.93	2.00	4.93	0.000202
	5	5.69	2.00	7.69	0.000493
	3	1.75	2.00	3.75	0.000117
	1	27.99	2.00	29.99	0.00750
Testing phase	13	- 0.42	2.00	2.42	0.0000488
	14	- 4.83	2.00	6.83	0.000389
	15	0.92	2.00	2.92	0.0000710
	16	19.84	2.00	21.84	0.00397

Table 8.5   Total-Risk with the actual process

Table 8.5 shows the results obtained with the actual process for assessing the Total-Risk value. It is clear to notice that the value for the Total-Risk (column 3) equals to 2%, does not change as in reality. This assumption causes that for all projects, there is a considerable difference between the current Total-Risk values and the Total-Risk (actual process) values. The point of this Table 8.5, is to show the wrong assumption that the Total-Risk is equally through all projects.

This difference creates a MSE value for each project and as a consequence, the total MSE value for the current process (0.0266 summation of column 5 in Table 8.5) is bigger than the MSE value for the NRAS(E) (0.005585 summation of column 5 Table 8.4). However, the MSE value for the testing set is smaller with the current process (0.00448 for project 13 to 16) than the NRAS(E) (0.005585 summation of the MSE values in column 5 for projects 13 to 16). In other words, the MSE value belonging to the actual process is bigger that the MSE value of the NRAS at the training phase; nevertheless, at the testing phase both values (actual process and the NRAS) are very similar. Perhaps with a bigger set of data for the training and testing sets, the results could be more accurate in determining a difference between the actual process and the NRAS.

The value of the current Total-Risk per project is changeable, is unique per project (Table 8.5, column 2). In reality, is not constant as it assumed with the actual process (Table 8.5, column 3). The BP10 (Figure 8.10) shows that approximates to the real values at the testing set, even do that the projects used at this phase were not used for training.

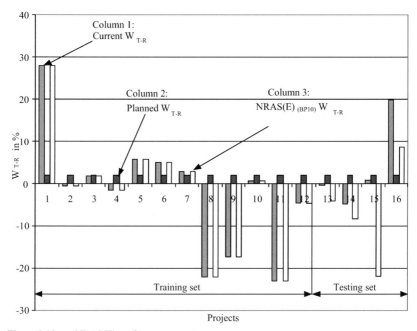

Figure 8.10   NRAS(E) performance

The meaning of the results shown in Figure 8.10 and Figure 8.11 is related to illustrate how well the NRAS imitates the Total-Risks (W_{T-R}) values per each project, obtained the results at the training set and at the testing set. Per each project in Figures 8.10 and 8.11, three results are shown, these are: the Planned Total-Risk (how the W_{T-R} was evaluated by the company), the Current Total-Risk (the real value of the W_{T-R} once the project finished) and the NRAS(E) Total-Risk (the value of the W_{T-R} obtained with the system). The difference of the results from Figure 8.10 and Figure 8.11 is that in Figure 8.10 the Total-Risk is in % and in Figure 8.11 the Total-Risk is in money terms.

Figure 8.10 shows a comparison of the results between the values of the current Total-Risk (current W_{T-R}, column 1 in each project), the planned values of the Total-Risk (planned W_{T-R}, column 2 in each project) using the current approach (see section 7.4, Formula 7.1) and the Total-Risk value obtained with the NRAS (W_{T-R} with NRAS(E)) (BP10, third column in each project).

As can be seen in Figure 8.10, the NRAS(E) is capable to imitate very well the current W_{T-R} in projects P1 to P12 (training set), in the rest of the projects (testing set), the NRAS(E) follows very closely the current W_{T-R} in projects 14 and close in projects 13 and 16; it is in project 15 where the NRAS(E) is not capable to imitate the current W_{T-R}

With the counterpart (the planned W_{T-R}), as observed in Figure 8.10. For each project, the planned W_{T-R} is assumed to be constant (2 %) and obviously exist a difference with the current W_{T-R} in each project. The results in Figure 8.10, shows the main contribution of the NRAS(E). Assuming that the Total-Risk value per each project is unique.

As shown in Figure 8.10, the values of the current W_{T-R} and the W_{T-R} with the NRAS(E) are the same from project 1 until project 12. This is under the training phase. With the planned W_{T-R}, the situation is different because the W_{T-R} is assumed to be constant (2 % of the project cost in this case); so this constant value differs greatly from the other two mentioned before (current W_{T-R} and W_{T-R} with NRAS(E).

From projects 13 until project 16, the situation between the current W_{T-R} and the W_{T-R} with the NRAS(E) is a bit different. Although they are not identical values, in 3 cases the NRAS(E) results follows the direction of the current W_{T-R} (for example, the current W_{T-R} is positive as happens with project 16, and the current W_{T-R} is negative with projects 13 and 14; in all cases the NRAS(E) imitates the same signs). This happens with values of the projects 13, 14 and 16. With project 15, there was a considerable difference between the current W_{T-R} and the W_{T-R} with the NRAS(E) values, because in this case the current W_{T-R} was positive and the value obtained with the NRAS(E) is negative.

The values for projects 13 to 16 obtained with the current process do not follow the current W_{T-R} or the NRAS(E) results. They continue to be constant and positive for each project. This is a serious disadvantage because in practice, the Total-Risk changes and is unique for each project.

As seen in Figure 8.10, the NRAS(E) results tend to follow the original decision made by the company towards the Total-Risk value per each of the 16 projects.

Figures 8.11 and 8.12, show the impact of the Total-Risk (W_{T-R} in €) value obtained in each of the three approaches: the current process values, planned process values and the NRAS(E). The goal is to show how well W_{T-R} was evaluated in each approach.

Figure 8.11 shows the W_{T-R} in money terms as isolated values for each project. Figure 8.11 shows the W_{T-R} values cumulated for each approach. In Figure 8.11 the values per project are obtained while multiplying the chosen W_{T-R} value (current, planned and NRAS(E)) per the AS netto (Table 8.2, column 6); For example, in the case of project 1, the current W_{T-R} value equals to 27.99 % (Table 8.3, column 7). Then, this value is multiplied by 926,546.27 € (AS netto of project 1, see Table 8.2, column 6). The result is: 0.2799 x 926,546.27 € = 259,340.00 €. Because project 1 is at the training phase, the same value is considered for the NRAS(E) $_{BP10}$ W_{T-R}. This is the reason why in Figure 8.11, the current W_{T-R} and the NRAS(E) $_{BP10}$ W_{T-R} are the same from project 1 until project 12.

For calculating the planned W_{T-R} value, it is only necessary to multiply the constant value of 2 % (Table 8.2, column 5) by the AS netto (Table 8.2, column 6). For example for project 1, this equals to: 0.02 x 926,546.27 € = 18,530.93 €.

In the case, for example a project which belongs to the testing set; the calculations are as follows (project 14 has been chosen):
Current W_{T-R} = -0.0483 x 4,203,315.20 € = - 203,020.12 €
Planned W_{T-R} = 0.02 x 4,203,315.20 € = 84,066.00 €
NRAS(E) $_{BP10}$ W_{T-R} = -0.0835 x 4,203,315.20 € = - 350,977.00 €

The same procedure is followed for the other projects used as testing set.

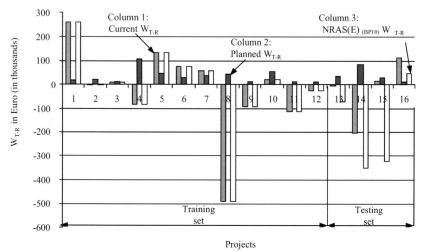

Figure 8.11   Total-Risk in money terms per project

The behavior of the values for the three approaches is similar as described before. The difference now is that the values are not any more a % of the project cost, now the values are money.

In this case, as was planned by the company. The firm has planned to obtain a profit in all the 16 projects. However, the reality was different; and a profit had occurred only in projects 1, 3, 5, 6, 7, 10, 15 and 16. In the other hand, in projects 2, 4, 8, 9, 11, 12, 13 and 14 the company had suffered a lost, in whatever way not so critical as occurred with projects 8, 11 and 14 where the loss is very dangerous for the company's financial security.

Although the values obtained with the NRAS(E) are not exactly the same as the actual ones, in 15 of the 16 projects the NRAS(E) follows similar paths (in orientation with the sign, positive or negative and some closeness with values) than the actual values. This is a great advantage because not only are the values of the NRAS(E) unique for each project, but also they are changeable either positive or negative. In relation with the values obtained with the actual process (Planned W_{T-R}), theses values are not reliable because are constant for all projects, without taking into account the changes that occurred with the current W_{T-R}.

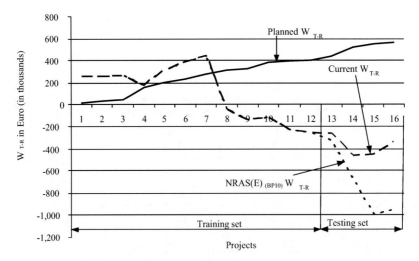

Figure 8.12   Cumulative Total-Risk per approach

The purpose and meaning of the results shown in Figure 8.12, are to describe the behaviour of the collective values of the W_{T-R} for the Planned W_{T-R}, the Current W_{T-R} and for the NRAS(E) W_{T-R}; taking into account all the 16 W_{T-R} values represented one per project. In other words, the results shown in Figure 8.12 are similar when the cash-flow of a project is built and represented, it is possible to observe that at project 16, the positive W_{T-R} value corresponds to the Planned W_{T-R}, while the negative values belong to the NRAS(E) W_{T-R} and the Current W_{T-R}. The process of how the values from Figure 8.12 were obtained, are shown by an example included in this section.

The cumulative Total-Risk per project, means the increasing value per project by additions (in this case, the additions represent the values of the W_{T-R} in Euro per project) in the three approaches: current W_{T-R}, with NRAS(E) and the planned W_{T-R}. For example at project 2, the cumulative summation is as follows:

Current W_{T-R} = 259,340.00 € $_{(project\ 1)}$ - 4,634.00 € (- 0.0047 x 985,921.37 €) $_{(project\ 2)}$
= 254,706.00 €

Planned W_{T-R} = 18,530.93 € $_{(project\ 1)}$ + (0.02 x 985,921.37) $_{(project\ 2)}$
= 18,530.93 € + 19,718.43 € = 38,249.36 €

NRAS(E) W_{T-R} = 259,340.00 € $_{(project\ 1)}$ - 4,634.00 € (- 0.0047 x 985,921.37 €) $_{(project\ 2)}$
= 254,706.00 €

Figure 8.12 shows the behavior of the W_{T-R} values within the 3 approaches. It is very important to notice firstly that the planned approach (the middle of Figure 8.12) follows always an incremental route, starts at 18,531.00 € and finishes at 562,183.00 €.

An opposite (comparing with the planned W_{T-R}, described above) situation happens with the current W_{T-R} values and the NRAS W_{T-R} values. Both start with gains (259,340.00 € and 259,340.00 €) and finish with losses (- 337,981.00 € and - 944,615.00 €). Obviously the values offered by the NRAS(E) are not identical as the values of the current W_{T-R} values; but the main point is that the NRAS(E) value (at project 16) can advise in this case to the company management board, that the projects (in this case 16), represent not such good opportunity for obtaining a final and attractive profit. The reality shows that at the end, the development of these projects will cause a considerable loss to the company. In the other side, the actual process (planned W_{T-R}, Figure 8.12) shows something different, shows that at the end, the 16 projects will achieve a win (at project 16, Figure 8.12), which is not true.

The NRAS has shown with its results, considerable advantages against the actual approach for assessing the Total-Risk. The NRAS represents an opportunity and opens a path for the contractor for managing project risks in a different way and predicting the possible cost of the Total-Risk.

8.5.5 Risk management evaluation

In this section, an evaluation of how well the risk was assessed in the testing phase is included. For doing this, the results of the BP10 (Table 8.4) from the NRAS(E) model were used and shown again in Table 8.6, plus information regarding the assessment of the risk for each of the 4 projects.

	Project	Current Total-Risk in %	Total-Risk (with ANN) in %	Difference with the current Total-Risk	Profit or Less	Risk was
Training phase	12	- 4.73	- 4.73	0	Profit	Well estimated
	10	0.72	0.72	0	Profit	Well estimated
	8	- 22.21	- 22.21	0	Profit	Well estimated
	6	4.98	4.98	0	Profit	Well estimated
	4	- 1.55	- 1.55	0	Profit	Well estimated
	2	- 0.47	- 0.47	0	Profit	Well estimated
	11	- 22.96	- 22.96	0	Profit	Well estimated
	9	- 17.35	- 17.35	0	Profit	Well estimated
	7	2.93	2.93	0	Profit	Well estimated
	5	5.69	5.69	0	Profit	Well estimated
	3	1.75	1.75	0	Profit	Well estimated
	1	27.99	27.99	0	Profit	Well estimated
Testing phase	13	- 0.42	- 4.10	0.0368	Profit	Well estimated
	14	- 4.83	- 8.35	0.0352	Profit	Well estimated
	15	0.92	- 21.90	0.2098	High profit	Very poorly estimated
	16	19.84	8.65	0.1119	Loss	Poorly estimated

Table 8.6   Risk management results with the NRAS(E)

The evaluation over the risk done by the NRAS(E) provides a very good basis for its results. In almost all projects, the risk was evaluated closely to reality; only in projects 15 and 16 there was a considerable discrepancy between the actual values and the NRAS(E) values.

As observed in Table 8.6, the risk related to all the projects used as testing set (1 to 12) was well evaluated. This is understandable and in some part obvious because the ANN knew the patterns already from the 12 projects of the testing set. However, for the remaining projects, in this case projects 13 to 16 which were used for testing the system. The results show that in 2 projects (projects 13 and 14) the risk was well estimated. With the remaining projects (15 and 16), the risk was very poorly and poorly estimated.

Table 8.7 shows the same type of information as in Table 8.6, but in this case corresponding to the actual approach.

Project	Current Total-Risk in %	Total-Risk (planned process) in %	Difference with the current Total-Risk	Profit or Loss	Risk was...
12	-4.73	2.00	6.73	Loss	Poorly estimated
10	0.72	2.00	1.28	Loss	Poorly estimated
8	-22.21	2.00	24.21	High loss	Poorly estimated
6	4.98	2.00	2.98	Loss	Poorly estimated
4	-1.55	2.00	3.55	Loss	Poorly estimated
2	-0.47	2.00	2.47	Loss	Poorly estimated
11	-22.96	2.00	24.96	High loss	Very poorly estimated
9	-17.35	2.00	19.35	High loss	Very poorly estimated
7	2.93	2.00	0.93	Profit	Well estimated
5	5.69	2.00	3.69	Loss	Poorly estimated
3	1.75	2.00	0.75	Profit	Well estimated
1	27.99	2.00	25.99	High loss	Very poorly estimated
13	-0.42	2.00	2.42	Loss	Poorly estimated
14	-4.83	2.00	6.83	Loss	Poorly estimated
15	0.92	2.00	1.08	Loss	Poorly estimated
16	19.84	2.00	17.84	High loss	Poorly estimated

Table 8.7   Risk management results with the actual process

The situation with the actual process in assessing the risk is quite different than the results obtained with the NRAS(E) as shown in Table 8.6.

From the set of projects used for training the NRAS(E), now with the actual approach, only in 2 projects (3 and 7) the risk was evaluated very well. With the rest of projects, (1, 2, 4, 5, 6, 8, 9, 10, 11 and 12) the risks in general were poorly estimated.

With the four projects used for testing the NRAS (E), now with the actual approach, in all cases the risk was poorly estimated. Even so, in project 16, the risk was very poorly estimated. As a comparison, with the NRAS(E), it was possible to estimate well the risks in 14 projects when with the actual approach only in 2 projects.

8.5.6 Conclusions

The managing of risks as was carried out through this chapter had left clear the meaning of risks in conditions that the risk value per each project is unique. In addition, another point which was mentioned is the one related to how the risk costs are planned (Planned W_{T-R}) and then what is the real risk cost (Current W_{T-R}). The NRAS has shown to be an alternative tool to assess the cost of risks (NRAS(E) W_{T-R}) in monetary terms.

It is a reality that the results offered by the NRAS are not 100% successful, the main reason of this situation is that the data available to train and to test the NRAS was very limited; Causing this limitation that the neural network of the NRAS was able only to recognize some of the patterns used at the testing phase from 12 projects. Even do, the system has proved that is possible to use neural networks to quantify in terms of money the cost of risks in construction projects.

The application example of the NRAS along this chapter was concentrated in 16 finished projects form one middle size German contractor. Based on the contractor's experience and in the German construction industry environment, a set of 17 Risk-Factors were developed. This application of the NRAS is not a limitation for the future used of the system. The NRAS is a flexible tool that can be modified and adapted to be used with even smaller or bigger contractors form developed or developing countries.

The results of the NRAS have shown that is possible to identify, analyse and quantify risks in terms of money. These results can considerably help the contractors at the bidding phase in assessing the possible value of the total risk (W_{T-R}). It is possible to observe how wrong were the risks evaluated (Planned W_{T-R}) by the contractor in almost all projects while comparing them with the current risks (Current W_{T-R}). In addition, the NRAS results (NRAS(E) W_{T-R}) imitate in an acceptable manner the current results (Current W_{T-R}). Even do, that this is a relative short example; it is clear that there are advantages in the NRAS(E) in compare with the actual process (Planned W_{T-R}) of assessing and forecasting the Total-Risk cost.

9   Conclusion

9.1   Conclusions

Three main conclusions can be drawn from the research. Firstly, it has been demonstrated that it is achievable to design, create, train and simulate an ANN, where this network is fed by Risk-Factors (17 which form the input layer) and as result, the ANN produces the value of the Total-Risk (which form the output layer) for the project evaluated. Secondly, even so that the training session of the ANN has been done with a very small set of data; it was attainable to see a pattern recognition when new projects were used at the simulation phase. Thirdly, it has been probe that the risk management tool, in this case the NRAS is capable to produce the Total-Risk value in % of the project cost for new projects.

In relation with the objectives of the present research work, it was possible to quantify the cost of the risks involved in construction projects with the creation, design and implementation of the NRAS. In addition, a list of Risk-Factors (RF) to model the behaviour of the risks for any given construction project was developed. It was possible to use artificial neural networks for identifying and evaluating risks in construction projects. These assumptions were made in chapter 1. In addition, the objective to offer the contractor a practical and useful risk management tool, ready to be used at the bidding phase of a construction project was achieved and as well the objective to develop a practical neuronal model for predicting the Total-Risk cost in construction projects. These objectives were made in chapter 7.

The main research contribution of this new approach is an alternative process to calculate in % of the project cost, the Total-Risk value for construction projects. The „Neuronal-Risk-Assessment System" is a complete flexible and adaptable approach that offers a different and new way of assessing risks, for the contractor's benefit. Artificial Neural Networks (ANN) formed one part of the model, their versatility, reliability and usefulness in other fields (medicine, electrical and mechanical engineering, aerospace engineering, etc), was and is a motivation for their implementation in the construction management specialisation.

9.2   Limitations

The main problem in using the NRAS is the one related to the data for training and testing the ANN. Having a small set of data, it really increase the chances to failed at the training and testing phase, because it would be difficult for the ANN to recognize the patterns.

After a series of discussions held with the experts of the Neuronal Research Group of the Technische Universitat Dresden (headed by Dr. Flach), it was clear that the data used to design, train and simulate the NRAS was minimal. Although the data was small, it was sufficient to test the system and acquire results. In order to gain more sound results, it is compulsory to achieve a higher quantity of data for proving the system in the industry. A problem existing at this research work was the possibility to get data from the contractor, perhaps this problem can be solved while developing the NRAS inside of a company and to use the product and results for the company and to only publish the development and behaviour of the NRAS.

One limitation present at this work is the one related with the knowledge necessary to work with ANN. The literature of ANN are not so easy to understand perhaps at the first try, it requires considerable amount of time and to be patient, especially when the reader will try to be self learner. After this, the researcher can start to try to use one of the available commercial software in ANN, which in the case of the Matlab programme has a very friendly environment to work (see section 5.8 from chapter 5).

9.3 General conclusions

The „Neuronal-Risk-Assessment System" is easy to use and manage (see sections 7.5, 7.6, 7.7, 7.8 and 7.9 in chapter 7 and sections 8.2, 8.3, 8.4, and 8.5 in chapter 8) because its structure is simple and is explained in detail in chapter 7. In comparison with some actual techniques of risk analysis like Monte Carlo Simulation (see section 2.7 and Table 2.1 in chapter 2, as well, see section 4.2.2, 4.2.3 and Table 4.1 in chapter 4) and Latin-Hypercube Simulation (see section 2.6 and Table 2.1 in chapter 2) as example, the effort to manage these techniques is less but the remuneration is also fewer; with the „Neuronal-Risk-Assessment System" the Total-Risk assessment (see sections 7.8 and in chapter 7 and section 8.3 in chapter 8) is done based on the evaluation of the most probable Risk-Factors (see section 7.7 in chapter 7) of the project in concern. In the other hand, with other risk analysis techniques, the risk assessment procedure demands knowledge of mathematical or probabilistic concepts (Monte Carlo Simulation and Latin-Hypercube for example) in which in many cases the project manager is not familiar with or can take considerable time when the application is done in situ in the industry.

The implementation of the Neuronal-Risk-Assessment System (see sections 7.5, 7.6, 7.7, 7.8 and 7.9 in chapter 7 and section 8.5 in chapter 8) as a risk management tool described in chapters 7 and 8, offers a different mode of assessing the possible risks in construction projects. This dissimilarity (distinction from the actual approaches to evaluate risks, Monte Carlo Simulation and Latin-Hypercube for example) is based on the single result obtained (W_{T-R}). It is possible to achieved a single value of the risk cost in money terms. This is complemented by providing to the project manager (project evaluator and risk manager) a friendly and easy way to assess the risks while using an evaluation form (Risk Evaluation Form, Figure 7.9). In addition, the NRAS is a human-intuition approach (experience of experts and project past data) which can be use as a tool for storing relevant data related not only to the project's risks.

One of the goals of the Neuronal-Risk-Assessment System is to act as an adviser to the company's management level (operational level), for forecasting the possible extra costs due to the Risk-Factors impact. As was observed with the results, the system was able to forecast the behaviour of the project's Total-Risk (W_{T-R}) in terms of money. Nevertheless, the Neuronal-Risk-Assessment System needs time to be formally implemented within a company's aims; especially with the departments dealing with the calculation of construction prices and project planning.

The Neuronal-Risk-Assessment System was designed to be implemented at the operational level (calculation phase). The reason is because at this phase is where the possible risks that the project might face are evaluated at the moment with a construction company. As well, decisions are known regarding the possibility of using insurance, partnerships, joint ventures, subcontractors; as measures to reduce, avoid, retain or share the risks. For that reason, knowing before hand the possible impact of the risks is a very valuable advantage.

9.4 Suggestions for further work

The NRAS is focused to appraise only one motive (perhaps the most relevant) which directly produces the project Total-Risk, the motive are the project Risk-Factors. Even do that, the relation between the Risk-Factors and Specific Risk-Factors (described in detail in section 7.7.3) with the project's Total-Risk is clear, logical and real. There are other sources of risk, which affect the project's Total-Risk like: the Project Manager and his team, the Client, etc. Figure 9.1 shows the graphical description.

These other sources are not handled with the NRAS, but their importance is not reduced or eliminated (see Risk-Factors 12 and 17 in section 7.7.2 and their corresponding Specific Risk-Factors in section 7.7.3, both in chapter 7). In other words, the research was concentrated on the project risks and relies on other researchers to investigate the impact of these other sources into the project.

158 Conclusion

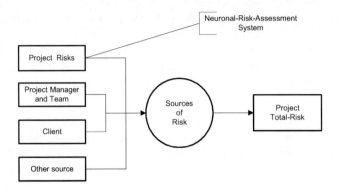

Figure 9.1   Project's Total-Risk and its causes

The sources of risk mentioned shown in Figure 9.1 (excepting Project Risks), perhaps can be investigated with other tools. For example, the impact of the project manager and team towards the project Total-Risk; maybe is possible to search this impact while using other risk management subjective methods like the Genesis of Management Confidence Technique described in section 4.3 from chapter 4. The client sources of risk, perhaps is possible to investigate while using the Estimating using risk analysis (ERA), method described in section 4.4 from chapter 4. It is out the scope of the present work to analysed these other risk causes, however, it is strongly believe on the importance of doing this extra research in order to compare the impact of each source in producing the project Total-Risk.

For the researchers interested in further developing of the NRAS, it is recommended to assure a considerable amount of data for both: the training and the simulations phases. Having this requirement covered, the possibility of obtaining acceptable and reliable results with the NRAS is increased. As well, a suggestion for further work with the NRAS is to investigate the use of different ANN algorithms. In this case, the backpropagation algorithm was employed, however other algorithms can also be introduce to the system and analyse their results and contributions to the risk management field.

Another suggestion, for example can be to try to use together ANN and Fuzzy logic with the system, and to analyse the advantages of doing this combination.

BIBLIOGRAPHY

Adler, M., Ziglio, E. (1996). Gazing into the Oracle: The Delpli Method and its Application to Social Policy and Public Health. Jessica Kingsley Publishers, London, UK.

Algorithmics (2000). The world's finest risk management solution for the enterprise, integrating market risk, credit risk, liquidity risk and limits. Algorithmics Incorporated, USA.

Aon Corporation (4th ed.) (1999). Enterprise risk management, part two. Aon Insights. The Aon Risk Services, Risk Management and Insurance Review, USA.

AS/NZS 4360 (1999). Risk Management. Standards Australia.

Baker, S., Ponniah, D., and Smith, S. (1998). Techniques for the analysis of risks in major projects. (p. 49, 567-572). Journal of the Operational Research Society, UK.

Bauch, U. (1994). Beitrag zur Risikobewertung von Bauprozessen. Dissertation. Technische Universität Dresden, Germany.

Beates, R.L. (1999). Glossary of Geology. Book News, Inc., Portland, USA.

Bishop, C. (1995). Neural Networks for Pattern Recognition. Clarendon Press, Oxford, UK.

Bossel, H. (1989). Simulation dynamischer Systeme. Vieweg Verlag, Braunschweig, Deutschland.

Boussabaine, A.H., Thomas, R., & Elhag, M.S. (1999). Modelling cost-flow forecasting for water pipeline projects using neural networks. Journal of Engineering, Construction and Architectural Management, (3, 213-224). Blackwell Science Ltd, UK.

Bhokha, S, & Ogunlana, S.O. (1999). Application of artificial neural network to forecast construction duration of buildings at the predesign stage. (p. 2, 133-144). Journal of Engineering, Construction and Architectural Management, Blackwell Science Ltd, UK.

CIRIA (2002). A simple guide to controlling risk. Construction Industry Research and Information Association. London, UK.

Conroy, G., and Soltan, H. (1998). ConSERV, a project specific risk management concept. (p. 353-366). International Journal of Project Management, Holland.

Cooper, D.F. (1999). Tutorial notes: The Australian and New Zealand standard on risk management. Broadleaf Capital International Pty Ltd.

Coy, D. (2000). Management immer wiederkehrender Projektrisiken. (p. 4, 22-29). Project management magazin, Germany.

CSA (1996). Risk Management: Guideline for Decision-Makers. Canadian Standards Association. Toronto, Canada.

Dress, G., Bahner, A. (1996). Kalkulation von Baupreisen, Wiesbaden und Berlin, Bauverlag, 4. Auflage.

Ernzen, J. J. and Schexnayder, C. (2000). One company's experience with design/Build: Labor Cost Risk and Profit Potential. (p. 126, 10-14). Journal of Construction Engineering and Management ASCE, USA.

Freund, D. (2000). Risk Management als Projektmanagement-Disziplin: Immer noch die " große Unbekannte". (p. 4, 52-58). Project management magazin, Germany.

Form, S., and Diederichs, M. (2000). Risikomanagement. Controlling Innovations Center GmbH&Co. KG. Dortmund, Germany.

Fowles J. (1978). Handbook of Futures Research. Greenwood Press, West Post Conn., USA.

Gibson, J. P. (1997).Contractual risk transfer strategies. International Risk Management Institute, Inc, USA.

Griffis, F.H., and Christodoulou S. (2000). Construction Risk Analysis for Determining Liquidated Damages Insurance Premiums: Case Study. (p. 407-413). Journal of Construction Engineering and Management, ASCE, USA.

Guthrie, V. H. and Walker, A. D. (2001). Enterprise Risk Management. ABS Group Inc., Risk & Reliability Division, USA.

Hasler, R. (2000). Alpha Project Line Risk Manager. (p. 4, 44-46). Project management magazin, Germany.

Haykin, S. (2nd ed.) (1991). Adaptive Filter Theory.Englewood Cliffs. NJ, Prentice Hall, USA.

Haykin, S. (1994). Neural Networks, a comprehensive Foundation. Macmillan College Publishing Company, Inc. USA.

Haykin, S. (2nd ed.) (1999). Neural Networks: A Comprehensive Foundation, Macmillan College Publishing Company, New York, USA.

Hicks, R.G., and Epps, J. A. (1999). Life Cycle Costs Analysis of Asphalt-Rubber Paving Materials. Final Report to Rubber Pavements Association. USA.

Himanen, V., Nijkamp, P., and Reggiani, A. (1998). Neural Networks in Transportation Applications. Ashgate Publishing Ltd, UK.

Holton, A. G. (1996). Enterprise Risk Management. Contingency Analysis, UK.

Hudson, F.V. (2000). Enterprise-wide Risk Management: Strategies for Linking Risk and Opportunity. Arthur Andersen, UK.

Huff, M. A., and Steven, G. (2001). Implementation of the Integrated Risk Assessment (IRA) Process Using the IRA Software. Schoolcraft; ABS Group Inc. Risk & Reliability Division (formerly JBF Associates, Inc); Knoxville, Tennessee, USA.

IPMA (International Project Management Association) (2000). International Advanced Training in Project Management. Copenhagen, Denmark.

Jaafari, M. (1987). Genesis of management confidence technique. (p. 3-60-80). Journal of Management in Engineering. ASCE, USA.

Javid, M., and Seneviratne, N. P. (2000). Investment Risk Analysis in Airport Parking Facility Development. (p. 298-305). Journal of Construction Engineering and Management, ASCE, USA.

Jordan, N. (1969). Some Thinking about 'System' in Themes in Speculative Psychology. Tavistock, London, UK.

Kähkönen, K., and Huovila, P. (1995). Risk Management of Construction Projects in Rusia, Proceeding of IPMA (International Project Management Asociation) International Symposium. (p. 237-241). The Russian Project Management Association Sovnet, St. Petersburg, Russia.

King, J. (1967). Economic Development Projects and Their Appraisal: Cases and Principles from the experience of the World Bank. The John Hopkins Press, USA.

Khosrowshahi, F. (1999). Neural network model for contractors' prequalification for local authority projects. (p. 3, 315-328). Journal of Engineering, Construction and Architectural Management Blackwell Science Ltd, UK.

Keitsch, D. (2000). Risikomanagement. Schäffer-Poeschel Verlag. Stuttgart, Germany.

Klein, J. H. and Cork, R. B. (1998). An approach to technical risk assessment. (p. 345-351). International Journal of Project Management, Holland.

Link, D. (1999). Risikobewertung von Bauprozessen. Dissertation, TU Wien, Austria.

Linsmeier, J. T., and Pearson, D. N. (1996). Risk Measurement: An Introduction to Value at Risk. University of Illinois at Urbana-Champaign, USA.

Nauck, D., Klawonn, F., and Kruse R. (1994). Neuronale Netze und Fuzzy-Systeme: Grundlagen des Konnektionismus, Neuronaler Fuzzy-Systeme und der Kopplung mit wissensbasierten Methoden. Vieweg, Wiesbaden, Deutsch.

McCallum, D., and Fredericks, I. (1997). Linking Risk Management and ISO 14000. MGMT Alliances Inc. And M+A Environmental Consultants.

Maccalis, J. (2000). Enterprise Risk Management: An Analytic Approach. Tillinghast Towers, UK.

Mak, S., and Picken, D. (2000). Using risk analysis to determine construction project contingencies. (p. 130-136). Journal of Construction Engineering and Management, ASCE, USA.

Maren, A., Harston, C. and Pap, R. (1990). Handbook of Neural Computing Applications. Academic Press Inc. San Diego, USA.

Matlab (2000). The Neural Network Toolbox. The MathWorks Inc., Natick, MA. USA.

Mawdesley, M., Askew, W., and O'Reilly, M. (1997). Planning and controlling construction projects: The Best Laid Plans...Addison Wesley Longman Limited. Essex. UK.

Medler, D. A. (1998). A Brief History of Connectionism. Biological Computation Project. Department of Psychology, University of Alberta, Canada.

Penter, Volker (1998). Einrichtung und Prüfung eines Risiko-Managementsystems. Vorlesung Risiko-Management-system. Arthur Andersen, UK.

Pilcher, R. (1992). Principles of Construction Management. Mc Graw-Hill Book Company, UK.

Perry, J.G., and Hayes, R.W. (1985). Risk and its management in construction projects. Proceeding. Institution of Civil Engineers Part 1: Design and Construction Jg 78/1985/Nr.3, S.499-521, UK.

Project Management Institute (1996). Project Risk Management. Upper Darby, PA: Project Management Institute, USA.

PwC Deutsche Revision (2000). Entwicklungstrends des Risikomanagements von Aktiengesellschaften in Deutschland-Empirische Studie des Instituts der Niedersächsischen Wirtschaft e.V., Hannover, Germany.

PwC Deutsche Revision (2000) . Finanzwirtschaftliches Risikomanagement deutscher industrie-und Handelsunternehmen. Fachverlag Moderne Wirtschaft. Frankfurt am Main, Germany.

Royer, P. S. (2000). Risk Management : The Undiscovered Dimension of Project Management. Upper Darby, PA: Project Management Institute, USA.

Rumelhart, D. E., Hinton, G. E., and Williams, R. J. (1986). Learning representations by backpropagation errors. Nature, 323, 533-536.

Sadeh, A., Dvir, D., and Shenhar, A. (2000). The Role of Contract Type in the Success of R&D Defense Projects Under Increasing Uncertainty. (p. 14-22) Project Management Journal, Holland

Schuppener, J., and Tillmann, W. (1999). KonTraG: Effects up on Loan business and credit rating. Credit practise, class 1999. (Nr.2, 20-23pp)

Site Safe (1999). Construction Safety Management Guide: Best Practice Guidelines in the Management of Health and Safety in Construction. Site Safe, New Zealand.

Shen, L. Y., and Wu, W. C. (2001). Risk assessment for construction joint veintures in China. (p. 76-81). Journal of Construction Engineering and Management, ASCE, USA.

Stockburger, D. W. (1996). Introductory Statistics: Concepts, Models and Applications. Southwest Missouri State University, USA.

The Building Research Association of New Zealand (BRANZ) (1999). Quantitative Assessment Methods for Determining Slope Stability Risk in the Building Industry. Study Report No.83. Riddolls & Grocott Ltd, New Zealand.

Thompson, P.A., and Perry, J.G. (1992). Engineering Construction Risks. Thomas Thelford London, UK.

Tsoukalas, L.H., and Uhrig R. E. (1996). Fuzzy and Neural Approaches in Engineering. John Wiley & Sons Inc, New York, USA.

Teji, T. (2000). Mastering risk beyond Turnbull: Risk Consulting. Arthur Andersen, UK.

Sudong, Y., and Tiong, L.K. R. (2000). NPV-at – risk method in infrastructure project investment evaluation. (p. 227-233). Journal of Construction Engineering and Management, ASCE, USA.

Wang, W. C., and Demsetz, L. A. (2000). Model for Evaluating Networks under Correlated Uncertainty-NETCOR. (p. 458-466). Journal of Construction Engineering and Management, ASCE, USA.

Wang, S. Q., and Tiong, R. L. K. (2000). Evaluation and management of political risks in Chinas`s bot project. (p. 126, 242-249). Journal of Construction Engineering and Management, ASCE, USA.

Wesley, H. J. (1997). Matlab Supplement to Fuzzy and Neural Approaches in Engineering. John Wiley & Sons, Inc, USA.

Widrow, B., and Stearns, S. D. (1985). Adaptive Signal Processing. Englewood Cliffs, NY: Prentice- Hall, USA.

Appendix I: Backpropagation example

The purpose of the following example, is to show how the Neuronal-Risk-Assessment System works while using the data from one project. The data from project 1 has been chosen to undertake the following sample.

a) Risk-Factors

The Risk-Factors assessed for project 1 are shown in figure 1. They represent the input data of the artificial neural network and will form its input vector.

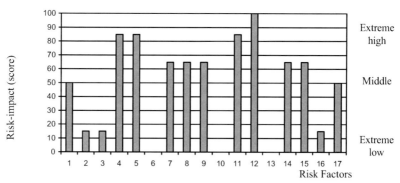

Figure 1   Risk-Factors scores of Project 1

b) Input vector

The same data shown in figure 1 is now arranged as a vector. The purpose is to prepare it for using as the input neurons of the artificial neural network.

$$
\begin{pmatrix} RF\text{-}1 \\ RF\text{-}2 \\ RF\text{-}3 \\ RF\text{-}4 \\ RF\text{-}5 \\ RF\text{-}6 \\ RF\text{-}7 \\ RF\text{-}8 \\ RF\text{-}9 \\ RF\text{-}10 \\ RF\text{-}11 \\ RF\text{-}12 \\ RF\text{-}13 \\ RF\text{-}14 \\ RF\text{-}15 \\ RF\text{-}16 \\ RF\text{-}17 \end{pmatrix} \quad \begin{pmatrix} 50 \\ 15 \\ 15 \\ 85 \\ 85 \\ 0 \\ 65 \\ 65 \\ 65 \\ 0 \\ 85 \\ 100 \\ 0 \\ 65 \\ 65 \\ 15 \\ 50 \end{pmatrix} \quad \begin{pmatrix} .50; \\ .15; \\ .15; \\ .85; \\ .85; \\ 0; \\ .65; \\ .65; \\ .65; \\ 0; \\ .85; \\ 1; \\ 0; \\ .65; \\ .65; \\ .15; \\ .50; \end{pmatrix}
$$

Vector elements   Vector scores   Vector in neuronal values

c) Uncertainty value

Formula 7.2 is used for obtaining the Total-Risk value (W_{T-R}), in this case corresponding to project 1. The details of formula 7.2, plus its purpose can be read in section 7.8.2.

$$W_{T-R} = \frac{P_{current}}{HSK_{TPC}} \cdot 100$$

$$W_{T-R} = \frac{359{,}000.00\ \text{€}}{1{,}285{,}900.00\ \text{€}} \cdot 100$$

$$W_{T-R} = 28.00\ \%$$

d) Output vector

The Actual (W+G) value, represents the output vector of the artificial neural network. The value of section e, is transformed in neuronal values.

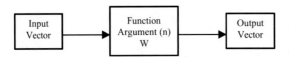

$\begin{bmatrix} \text{Actual} \\ \text{(W+G)} \end{bmatrix}$ $\begin{bmatrix} 28.00 \end{bmatrix}$ $\begin{bmatrix} .28; \end{bmatrix}$

Vector elements   Vector scores   Vector in neuronal values

e) Artificial Neural Network Design

Figure 2 shows the conceptual design of the neural network. Refer to section 7.10 for a detailed explanation of the inputs and outputs model representation, and as well for an explanation of the argument of the transfer function and the weigths.

Input Vector → Function Argument (n) W → Output Vector

Figure 2   Neuronal Design

f) Artificial Neural Network Structure

Based on the work done until now, the structure of the network is shown in figure 3. The input, hidden and output neurons are shown and also the conncetion weigths.

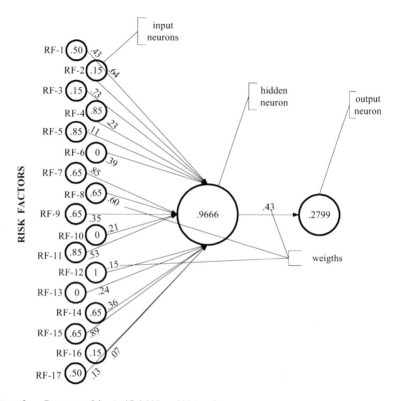

Figure 3 Structure of the Artificial Neural Network

NOTE: For the details regarding the steps of applying the backprogation algorithm (Forward and Backward propagations), refer please to section 5.7.

g) Forward Propagation

Step 1: Calculation of the outputs to the hidden neuron

This is done by multiplying the Risk-Factor scores for their corresponding weights which connect to the hidden neuron. The outputs are shown in table 1, column 4.

Risk-Factor	Risk-Factor Score	Risk-Factor Weigth	Outputs to hidden neuron
RF-1	0.50	0.43	0.21
RF-2	0.15	0.64	0.09
RF-3	0.15	0.73	0.10
RF-4	0.85	0.23	0.19
RF-5	0.85	0.11	0.09
RF-6	0	0.39	0
RF-7	0.65	0.85	0.55
RF-8	0.65	0.60	0.39
RF-9	0.65	0.35	0.22
RF-10	0	0.21	0
RF-11	0.85	0.53	0.45
RF-12	1	0.15	0.15
RF-13	0	0.24	0
RF-14	0.65	0.36	0.23
RF-15	0.65	0.89	0.57
RF-16	0.15	0.07	0.01
RF-17	0.50	0.13	0.06

Summation = 3.31

Table 1   Outputs to hidden neuron

h) Forward Propagation
Step 2: Output from the hidden neuron

While using the transfer function (log-sigmoid), the summation of the inputs coming from the input neurons is used in order to obtain the output of the hidden neruon which will connect to the output neron.

$$\text{Output} = \frac{1}{1+e^{(-\text{sum})}}$$

$$\text{Output} = \frac{1}{1+e^{(-3.31)}} = 0.96$$

i) Forward Propagation
Step 3: Input to the output neuron

The input to the output neuron is calculated by multiplying the output of the hidden neuron by its corresponding weight.

$$\text{Output} = 0.96 \cdot 0.43 = 0.41$$

j) Forward Propagation
Step 4: Error (Difference)

The value obtained in step 3, is called the „actual value" which is compare with the desired value; in this case .28. The error is the difference between them.

Actual value   = 0.41
Desired value   = 0.28
Error (Difference) = 0.13

k) Back Propagation
Step 5: Adjusting weight for the hidden neuron

This is done by multiplying the output of the hidden neuron by the Error (step 4) and by the learning rate (0.25 proposed).

Adjusting weight = $0.96 \cdot 0.13 \cdot 0.25 = 0.03$

l) Back Propagation
Step 6: New weight for the hidden neuron

The adjusting weight (step 5) is added to the old weight of the hidden neuron.

New weight = $0.43 + 0.03 = 0.46$

m) Back Propagation
Step 7: Error in the hidden neuron

To obtain the error in the hidden neuron, the Error (step 4) is multiplyied by the old weight.

Error in the hidden neuron = $0.13 \cdot 0.43 = 0.05$

n) Back Propagation
Step 8: Adjusting weights for the input neurons.

These adjusting weights are calculated by multiplying the Risk-Factor score by the error in the hidden layer (step 7) and by the learning rate (0.13 proposed). See column 5 in table 2.

170 Appendix I: Backpropagation example

Risk-Factor	Risk-Factor Score	Error in hidden neuron	Learning rate	Adjusting weigths
RF-1	0.50	0.05	0.13	0.003
RF-2	0.15	0.05	0.13	0.001
RF-3	0.15	0.05	0.13	0.001
RF-4	0.85	0.05	0.13	0.005
RF-5	0.85	0.05	0.13	0.005
RF-6	0	0.05	0.13	0
RF-7	0.65	0.05	0.13	0.004
RF-8	0.65	0.05	0.13	0.004
RF-9	0.65	0.05	0.13	0.004
RF-10	0	0.05	0.13	0
RF-11	0.85	0.05	0.13	0.005
RF-12	1	0.05	0.13	0.006
RF-13	0	0.05	0.13	0
RF-14	0.65	0.05	0.13	0.004
RF-15	0.65	0.05	0.13	0.004
RF-16	0.15	0.05	0.13	0.001
RF-17	0.50	0.05	0.13	0.003

Table 2   Adjusting weights for the input neurons

o) **Back Propagation**
   Step 9: New weights for the input neurons

Finally, the new weights (table 3, column 4) for each input neuron are calculated. This is made by adding to each Risk-factor weigth (table 3, column 2) its corresponding adjusting weight (table 2, column 5).

Risk-Factor	Risk-Factor Weigth	Adjusting weigths	New Weights
RF-1	0.43	0.003	0.433
RF-2	0.64	0.001	0.641
RF-3	0.73	0.001	0.731
RF-4	0.23	0.005	0.235
RF-5	0.11	0.005	0.115
RF-6	0.39	0	0.39
RF-7	0.85	0.004	0.854
RF-8	0.60	0.004	0.604
RF-9	0.35	0.004	0.354
RF-10	0.21	0	0.21
RF-11	0.53	0.005	0.535
RF-12	0.15	0.006	0.156
RF-13	0.24	0	0.24
RF-14	0.36	0.004	0.364
RF-15	0.89	0.004	0.894
RF-16	0.07	0.001	0.071
RF-17	0.13	0.003	0.133

Table 3   New weights for the input neurons

Until now, the Forward and Back propagation for one epoch or cicle is completed. The same procedure is followed by the network until the error is zero or near.

Appendix II: Simulation models results

a) Simulation results from model NRAS(B) to NRAS(J)

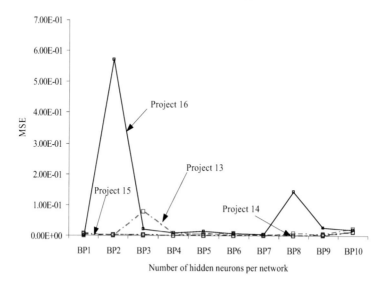

Model NRAS(B)

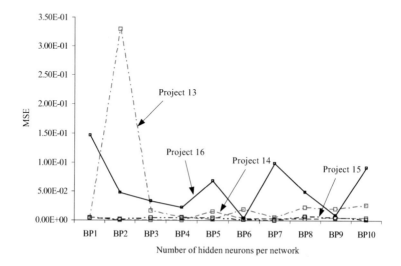

Model NRAS(C)

174   Appendix II: Simulation results from model NRAS(B) to NRAS(J)

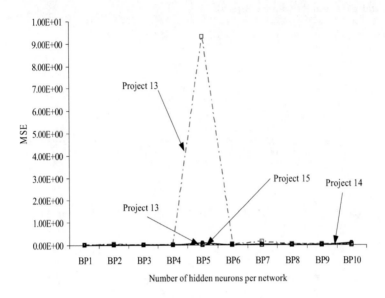

Model NRAS(D)

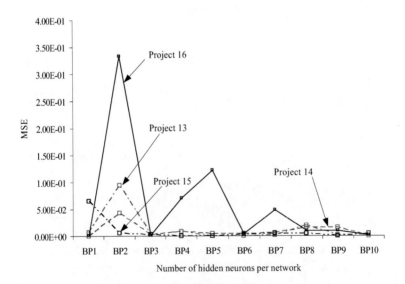

Model NRAS(E)

Appendix II: Simulation results from model NRAS(B) to NRAS(J)   175

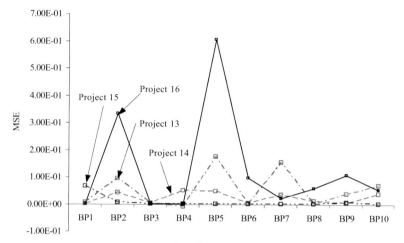

Model NRAS(F)

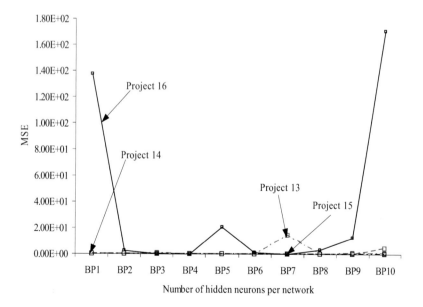

Model NRAS(G)

176 Appendix II: Simulation results from model NRAS(B) to NRAS(J)

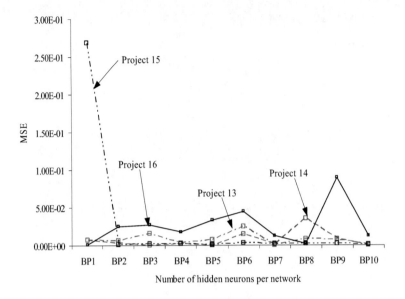

Model NRAS(H)

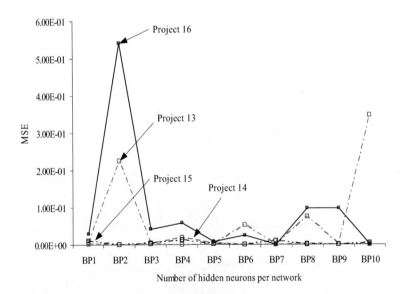

Model NRAS(I)

Appendix II: Simulation results from model NRAS(B) to NRAS(J) 177

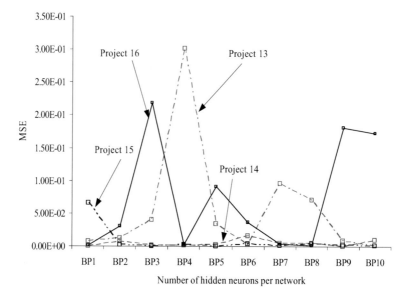

Model NRAS(J)

b) Backpropagation networks performance at the training phase

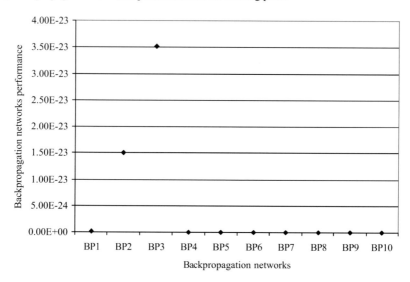

Figure 1   Backpropagation network performance for model NRAS(B)

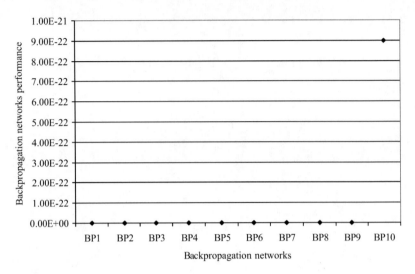

Figure 2   Backpropagation network performance for model NRAS(C)

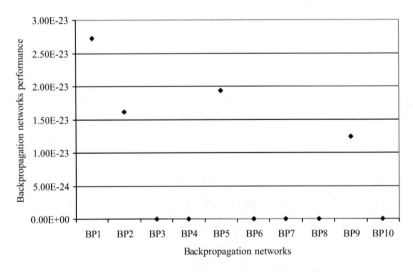

Figure 3   Backpropagation network performance for model NRAS(D)

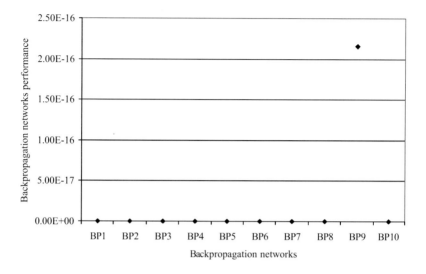

Figure 4　　Backpropagation network performance for model NRAS(E)

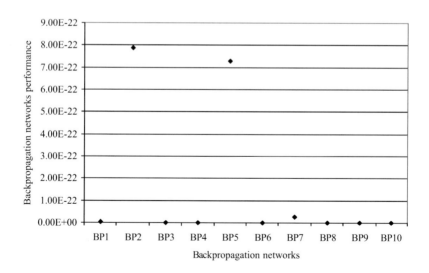

Figure 5　　Backpropagation network performance for model NRAS(F)

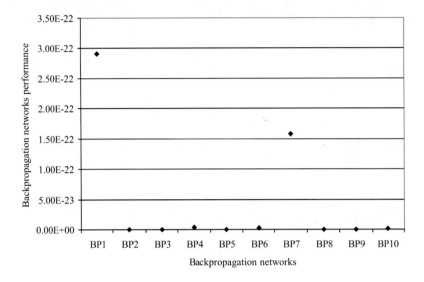

Figure 6 Backpropagation network performance for model NRAS(G)

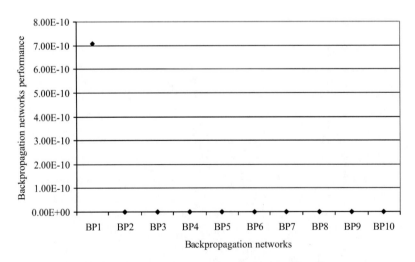

Figure 7 Backpropagation network performance for model NRAS(H)

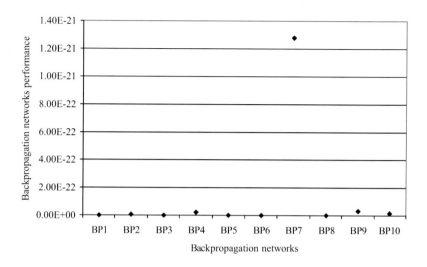

Figure 8   Backpropagation network performance for model NRAS(I)

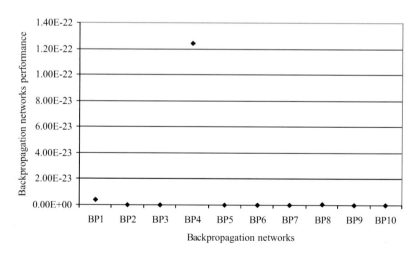

Figure 9   Backpropagation network performance for model NRAS(J)